Community-Driven Regulation

Urban and Industrial Environments
Series editor: Robert Gottlieb, Henry R. Luce Professor of Urban and Environmental Policy, Occidental College

Community-Driven Regulation

Balancing Development and the Environment
in Vietnam

Dara O'Rourke

The MIT Press
Cambridge, Massachusetts
London, England

This book was set in Sabon by Graphic Composition, Inc.
Printed on recycled paper and bound in the United States of America.

Library of Congress Cataloging-in-Publication Data

O'Rourke, Dara.
  Community-driven regulation : balancing development and the environment in Vietnam / Dara O'Rourke.
      p. cm.—(Urban and industrial environments)
  Includes bibliographical references and index.
  ISBN 0-262-15108-1 (hc. : alk. paper)—ISBN 0-262-65064-9 (pbk. : alk. paper)
      1. Sustainable development—Vietnam. 2. Community development—Vietnam. 3. Vietnam—Economic policy—1975– I. Title. II. Series.
  HC444.Z9E56   2004
  338.9597′07—dc21   2003046300

10   9   8   7   6   5   4   3   2   1

Contents

Preface

Stooped over, knee-deep in the rich soils and delta waters of Vietnam—from which farmers have been coaxing rice for over a thousand years—Nguyen Thi Hoa has an unlikely front-row seat for one of the fastest processes of industrial development in the world. From the rapidly shrinking rice paddies around Bien Hoa city, Ms. Nguyen and her neighbors have watched as Vietnam has opened up to the global economy, and foreign multinational corporations such as Nike, Fujitsu, and Sanyo have established factories in Dong Nai province. She has marveled at the growth of large state-owned enterprises throughout the province, including a pulp and paper mill a stone's throw from her home. And she has watched as markets and even minimarts have sprung up around town, selling products never before seen in Vietnam. Rationing is now a remnant of the past as people can afford meat, televisions, and even motorcycles—indicators more telling for Ms. Nguyen than GDP growth rates or foreign exchange earnings.

But Ms. Nguyen and others have also seen the adverse impacts this development is bringing to Vietnam. They have watched small farms and worker cooperatives—the centerpieces of the Communist economic model—be bought out or shut down and replaced by private agribusiness firms. They suffer declining yields from their fruit trees, their animals become sick from drinking now-polluted groundwater, and children and older people suffer respiratory ailments and headaches when the wind carries dirty air from nearby factories.

Throughout my research in Vietnam, villagers and urbanites, farmers and factory workers, retirees and college students, told mixed stories of the impacts of development in Vietnam—a development process they generally supported. People welcomed Vietnam's entry into the global economy with the modernization of infrastructure, institutions, and industry that has emerged. Yet, they were simultaneously concerned about damage to local environments, workers' health, Vietnamese society, and about who was really benefiting and who was paying the costs of this development. Rural villagers wondered out loud whether they had won battles against imperialism and capitalism in the 1950s, 1960s, and 1970s, only now to lose a bloodless war against a new global force.

And everywhere I traveled in Vietnam, I was asked whether there was not a better way—a way for Vietnam to develop, modernize, and industrialize, without abandoning its commitment to rural farmers, to social equity, to workers' and women's rights, and to protecting the environment. Often the question posed was whether Vietnam could learn from the United States or Europe or other newly industrialized countries for strategies to avoid the negative impacts of industrial development. But there are no easy answers or models to apply to Vietnam. Industrialization in the United States, Europe, Japan, Korea, and Taiwan involved increased pollution, depleted natural resources, and exploited workers.

But these questions and concerns of Vietnamese farmers, workers, and urbanites, and efforts in Vietnam to balance recent development with environmental protections and social equity, stuck with me, and in many ways motivated this study. Working in Vietnam also helped me to see that foreign models and theories are often not the answer, as there is much to learn from looking inward at processes already underway in Vietnam.

This challenge of balancing development and the environment obviously has wide relevance beyond Vietnam. Virtually every country in the world struggles with the question of how to promote industry, create jobs, and modernize, while minimizing or preventing adverse impacts to workers, communities, and the environment. The experience of Vietnam offers a particularly interesting lens through which to examine these questions and challenges. Vietnam was largely excluded from the global economy until the early 1990s. However, in just the last ten years, Vietnam has emerged as one of the new "Tiger" economies of Asia, experiencing phenomenal economic growth. With the lifting of the U.S. trade embargo in 1994 and the entry of the World Bank, Asian Development Bank, International Monetary Fund, and a flood of bilateral donors, aid agencies, and trade agreements, Vietnam has very rapidly joined the global economy.

Vietnam thus presents an excellent case study for examining how a country makes the transition to industrial development and global economic integration, while facing the challenges and opportunities of protecting workers and the environment. Vietnam offers an almost quintessential case for analyzing trade-offs between development and the environment, and for analyzing policy options and strategies for minimizing adverse impacts of industrial development.

Vietnam also provides a window on the effectiveness of a range of state policies for environmental protection. As is clear from examples around the world, environmental regulations often fail to achieve their stated goals, particularly during periods of rapid industrial development. Existing theories, however, fail to adequately explain the underlying social and political processes that lead to failures of environmental regulation, and more importantly, to explain mechanisms and processes that can motivate state and civil society actors to effectively control or prevent degrading development. With the rapid growth of domestic and foreign aid programs, Vietnam provides examples of efforts to strengthen state agency capacities and policies, and to employ market mechanisms to better balance development and the environment.

Many of the crucial dynamics, however, occur under the radar of these sweeping perspectives. To understand the tensions around development and the environment in Vietnam, one must understand the worlds of common citizens such as Ms. Nguyen, who struggled, collectively, to demand environmental and workplace protections. One of the surprises of this research was that pollution was regulated effectively at any time. In situations where both the existing literature and common sense would expect failure—where there were weak government capacities, clear incentives to promote industry over the environment, and firms more powerful than communities—I found successful cases of regulation. In a variety of contexts and types of firms, environmental regulations were implemented against rather severe odds. These cases drove me to seek to understand the conditions under which regulation can be effective, and industrial activities can be balanced with environmental and social concerns.

This book uses six cases of industrial development and environmental impacts in Vietnam to advance an admittedly preliminary framework for thinking about strategies for environmental regulation. This framework, which I call Community-Driven Regulation (CDR), helps to specify processes that influence state and firm responses to pollution problems. In successful cases of CDR, community actions play a central role in pressuring state agencies to improve their monitoring and enforcement of environmental regulations. Community mobilizations essentially begin a dialog between impacted stakeholders and the state, leading to debate, conflict, and sometimes bargaining over developmental and environmental trade-offs. This pressure and dialog can also actually strengthen state capacities and commitments for enforcement.

Examples of CDR display a dynamic of mutually reinforcing interactions—sometimes collaborative, sometimes confrontational—between organized communities and state environmental agencies, pointing to the potential for "state–society synergy" and the coproduction of regulation. Through this process, Vietnam is essentially evolving a system of regulation driven by mobilized communities and their complaints, to which provincial government agencies are increasingly responsive. Nonetheless, to date, these dynamics have remained informal and have not been codified in new laws or regulatory processes. Research results thus also can be used to argue that future policies can strengthen both state and community roles in environmental protection and build on existing dynamics to promote more ecologically sound development.

Research

Research in Vietnam sometimes seems to require, above all else, the ability to sit for many hours sipping green tea and listening. As conversations expand from small talk, to gently brushing the edges of a topic, to directly engaging in a controversial question, the tea leaves expand as they are reheated over and over with water from a ubiquitous red thermos, sucking the last bits of flavor out of each leaf. It is almost as if during the process of the water losing its heat, and the tea gaining in bitterness, community members finally give in to the pleas of an outsider attempting to understand some aspects of Vietnam.

It was on an afternoon of drinking green tea, and later a jet black herbal health drink, that I began to get a sense of the scale of adverse impacts of Vietnam's transition process on local communities, workers, and the environment. On that day, community members complained of air pollution that had been blowing into their houses from a nearby factory and causing a range of health problems. With a little bit of trust and a lot of tea, people began to detail their concerns about the environmental and health impacts of recent development.

Through a targeted sampling process I interviewed the actors involved in specific cases of pollution controversies in Vietnam. I interviewed community members about their perspectives on local industry and pollution impacts in their neighborhoods, their opinions on the effectiveness of government regulation, and what actions they had taken themselves regarding these issues. I interviewed factory managers to understand their perspectives on the environmental impacts of their activities, potentials for pollution reductions, internal opportunities and constraints on production changes to reduce pollution, and their relations with state agencies and the community. And I interviewed the government inspectors responsible for regulating these factories, and the government officials responsible for promoting economic development in these provinces. Throughout the case studies, I also engaged and interviewed nonstate actors such as nongovernmental organizations (NGOs), reporters, and international advisors and consultants.

This book seeks to piece together and examine these different perspectives and interactions between state agencies, firms, and communities around cases of industrial pollution. Through a comparative analysis of national policies and local dynamics, the book analyzes processes through which community members, state actors, and ultimately factory managers take actions to reduce pollution. Six detailed case studies—including state-owned chemical plants, a Nike shoe factory, a Taiwanese textile plant, and a state-owned pulp and paper mill—are comparatively assessed throughout the book.

The research for this book involved thirteen trips to Vietnam between 1994 and 2000, twelve months of fieldwork, over 180 interviews, and site visits to thirty factories and their surrounding communities. The research included both macrolevel analysis of state actions and policies, and microlevel research inside the six factories and communities. Along with the field-work, I conducted analysis of media reports related to the six cases as well as to developmental and environmental issues more broadly related to Vietnam. I reviewed a wide range of government documents on pollution incidents, regulatory actions, inspections, fines, Environmental Impact As-sessments (EIAs), and public complaints. And I was given unparalleled ac-cess to government reports through two United Nations projects and a World Bank project with which I was affiliated during the course of the re-search. The U.N. projects allowed me to conduct research inside provincial-level environmental agencies in Dong Nai and Phu Tho provinces. I also interviewed a wide range of national government environmental officials and international agency staff.

My research strategy however, was not to aspire to the impossible image of an unbiased anthropologist peering out from a tent and silently observ-ing the undisturbed functioning of Vietnamese society. To some degree, I became a part of the cases I analyzed. My questions and interactions no doubt influenced the people I interviewed. As I tried to minimize my in-fluence on informants, throughout the research I also tried to stay aware of the influences and shadows I cast as I entered into peoples' lives. Nonetheless, some of my most important learning came from interacting with different stakeholders, engaging their positions and perspectives, and seeing how they responded.

The Nike case is the clearest example of this interaction as my research impacted both the firm and a range of nongovernmental actors pressuring the firm. In seeking to understand what motivates a firm to change, I was fortunate enough to participate directly in dialogs between Nike, Tae Kwang Vina—its Korean contractor in Vietnam, workers in the plant, NGOs in Vietnam and the United States, and the media. This direct par-ticipation helped me to understand the processes and dynamics at play in the controversy around this specific factory, and provided a window on larger debates about globalization and global firms such as Nike. To lesser degrees in the other cases, I engaged the institutions and actors involved and my understanding subsequently deepened.

Acknowledgments

In many ways this book has been a community-driven process. Although I am ultimately responsible for the work, my research and thinking were continually supported, challenged, and strengthened by a diverse community of scholars, practitioners, activists, and community members. Although it would be impossible to name every person who helped along the way, a number of individuals deserve special mention.

First, I want to thank Clay Morgan, my editor at the MIT Press for supporting and shepherding this project. I also want to thank five anonymous reviewers for constructive advice that has made the book much stronger.

This book would not have been completed without the sustained support, advice, and gentle prodding of Peter Evans. Peter was the quintessential advisor, always pushing me to ask tough questions and to make bold claims. Rachel Schurman and Tim Duane also provided enormous support and feedback throughout the preparation, research, and writing of the dissertation that preceded this book. My colleagues and friends in the Energy and Resources Group at UC Berkeley and the Department of Urban Studies and Planning at MIT also significantly influenced my thinking and arguments over the course of the project: William Boyd, Scott Prudham, Thomas Sikor, Navroz Dubash, and Paul Sabin instigated discussions and debates that fueled the work. Archon Fung has been amazingly generous with his time and critical thinking over the last decade and has helped me to better understand both theory and practice. Richard Locke and Judith Tendler provided generous input and support during the critical editing phases of the book at MIT.

I could not have conducted the research for this book without extensive and long-term support of researchers in Vietnam. Dr. Dinh Van Sam and Dr. Tran Van Nhan from the Institute for Environmental Science and Technology (INEST) at the Hanoi University of Technology sponsored and supported my work from the summer of 1994 until its conclusion. Ms. To Thi Thuy Hang and Ms. Nguyen Ngoc Ly from the United Nations Development Programme offered insights, support, and friendship throughout my time in Vietnam. I benefited as well from the excellent research assistance of Ms. Tang Thi Hong Loan, Mr. Pham Huy Han, Mr. Le Van

Lang, Mr. Mike Hooper, and Mr. Eungkyoon Lee. My understanding of Vietnam was also deepened by interactions with Dr. Nguyen Trung Viet, Dr. Bach Tan Sinh, Dr. Le Thac Can, Mr. James MacNeil, Dr. Anders Thuren, and Dr. Jeremy Carew-Reid. I also want to thank Dr. Rob Koudstaal, Dr. Jack Day, Dr. Skip Luken, and Mr. Nguyen Khac Tiep for providing the opportunity to participate in the UNIDO projects that enriched my research. I must of course also acknowledge and thank the many villagers, workers, factory managers, and government officials in Vietnam who gave generously of their time and insight. Without their courage and actions I would find much less hope for the potential to balance development and the environment in Vietnam.

The research was made possible by generous support from a National Science Foundation Graduate Fellowship, Foreign Language and Area Studies Fellowship, Switzer Fellowship, Institute for International Studies Simpson Memorial Research Fellowship, the Center for Southeast Asian Studies and the Energy and Resources Group at UC Berkeley, and the Department of Urban Studies and Planning at MIT.

Not surprisingly, my family has played an instrumental role in the journey that brought me to the finish line of this marathon. My brother, Gerry, provided a couch to sleep on and much support on my stopovers in Hong Kong. My brother, Dan, provided a sense of humor and perspective on the whole project (including a "suspicious package from Nike" halfway through the research). My mother and father provided unfailing encouragement for my academic and activist pursuits, and instilled a deep desire to make a difference in the world. Finally, I want to thank Cathy Cha for her love and support and laughter throughout this process. Without her companionship, encouragement, intellectual support, and reminders that I was almost done, the process would have been much less meaningful.

Acronyms and Vietnamese Phrases

Acronyms

ADB	Asian Development Bank
CDR	Community-Driven Regulation
CEM	Center for Environmental Management
CP	Communist Party
DOI	Department of Industry
DOSTE	Department of Science, Technology, and Environment
DSEE	Division of Science, Education, and Environment (within the Ministry of Planning and Investment)
EIA	Environmental Impact Assessment
EMD	Environmental Management Division (Hanoi)
FDI	Foreign Direct Investment
GDP	Gross Domestic Product
IMF	International Monetary Fund
JV	Joint Venture
LEP	Law on Environmental Protection
MNC	Multinational Corporation
MOH	Ministry of Health
MOI	Ministry of Industry
MOSTE	Ministry of Science, Technology, and Environment
MPI	Ministry of Planning and Investment
NEA	National Environment Agency

NGO	Non-Governmental Organization
ODP	Overseas Development Assistance
PC	People's Committee
SOE	State-Owned Enterprise
UNDP	United Nations Development Program
UNIDO	United Nations Industrial Development Organization
VNCPC	Vietnam Cleaner Production Center
WTO	World Trade Organization

Vietnamese Phrases

Doi Moi	Vietnam's political and economic reform program; literally, "renovation"
khu pho	Area or zone—a group of streets in a neighborhood
lao dong	Labor
ngoai tich	Outsider
noi tich	Insider
phuong	Ward or district
thanh nien	Youth
thon	Hamlet
tinh	Province
xa	Commune

1

The Challenges of Balancing Development and the Environment

1.1 Introduction

Countries seeking to advance "sustainable development," and even those interested simply in sustaining their development, face challenges that are both daunting and sometimes contradictory: how to both promote economic development and protect the environment upon which development depends. This challenge can hardly be overstated. Many developed and developing countries alike have tried and failed to strike a balance between growth and the environment. Grand theories and pragmatic policies have failed to provide clear signposts for a path to sustainable development. And countries that have ignored environmental and social concerns in their quests for growth, development, and jobs, have often ended up paying later in mounting pollution, worsening health impacts, and increasing costs of environmental protection.

A range of theories do, of course, exist for balancing development and the environment, and more specifically for controlling environmental degradation. Government policies, market mechanisms, and civil society actors have all been held up as central to striking this balance. However, more questions than answers remain in actually advancing sustainable development.

For instance, government regulation can and does play a role in controlling the worst forms of environmental degradation. But as is now clear, regulation of economic activities requires state capacity and commitments that are all too rare in developing countries. Market mechanisms also hold promise to motivate more environmentally sound actions. However, few developing countries have the institutional conditions necessary to advance

market-based strategies for environmental protection. And public participation certainly appears to be an engine for demanding environmental and
health protections during development; however, few countries have effective processes for public participation in environmental disputes. When
we look around the world, few governments offer guides to sustainability.

Is there then little hope for balancing environmental protections during
the drive for economic development? Are developing countries doomed to
repeat the mistakes of industrialized countries—polluting their environments and depleting their natural resources as they industrialize? Are the
world's poorest countries trapped in this trade-off, forced to accept degradation of their environments and increased burdens on human health in
the pursuit of economic development?

The difficulties of effectively regulating industry during development are
certainly significant. However, evidence shows that there are mechanisms
that can support governments in their efforts to enforce environmental
standards. Regulation can work, even against the difficult odds faced by
developing country governments. Real-world cases provide hope for promoting more balanced and more sustainable development.

This book examines the challenges of balancing development and the
environment, and identifies processes that hold potential for supporting
the "coproduction" of regulation, with communities and state agencies
working together to balance economic and environmental goals. Government policies, market dynamics, and civil society participation are all critical for advancing this balance. In this model, effective regulation involves
a combination of bottom-up pressures—essentially the energies and actions of average community members—and the responses of front-line environmental agencies. Community actions stand at the heart of a process
that can drive state environmental agencies to take actions to pressure
firms to reduce pollution. I call this dynamic "Community-Driven Regulation (CDR)." These interactions between the state, communities, and
firms are critical to the implementation of environmental regulations. Understanding the conditions under which communities drive these processes
and state authorities and firms respond is the central inquiry of this book.

To explain these dynamics, I draw from evidence gathered through six
years of research in Vietnam. The dynamics of Vietnam are relevant to the
many countries experiencing similar struggles and challenges of promot-

ing economic development (and global economic integration) while trying to protect local cultures and environments. Vietnam presents a difficult case for examining these challenges. The country has experienced a massive inflow of foreign investors, is among the fastest industrializing countries in the world, and at the same time, has weak state regulatory agencies. It also experiences problems with corruption in state enforcement, and continues to support highly polluting state-owned enterprises. And yet even under these constraints we find cases of effective regulation and cases that suggest the potential for balancing development and the environment.

Consider for example the case of a Taiwanese textile factory, the Dona Bochang Textiles company, operating in Vietnam and selling its products around the world. A towel hanging behind the manager's desk bearing a human-sized likeness of Mickey Mouse explains in a single image the factory's aspiring role in the global economy. Just beyond Mickey, through the manager's window is a factory full of young Vietnamese women (recently migrated from rural farms), working in one of several hubs of this growing Taiwanese multinational supply chain, producing cross-branded goods to the rest of the Asia.

The Dona Bochang factory and the industrial system it represents appear not only to be on the cutting edge of global production, but also an almost completely unregulatable element of that system. The firm is a profitable company with connections to large brand names, but without a reputation of its own that can be publicly damaged. No activist or consumer campaigner in the world has ever heard of Dona Bochang, or could even find this factory in the complex of the global production system. The company has set up operations in partnership with local government officials. The Communist Party (CP) of Dong Nai Province actually owns 12.5 percent of the factory. There are no local nongovernmental organizations (NGOs) to put pressure on the firm. And to top it all, the local regulatory agency— which is both young and weak—answers to the very same Party officials who own a portion of the factory. How could any government agency, let alone a weak one, regulate a multinational corporation such as this as it plays off government agencies, and even pits countries against one another?

The Dona Bochang textiles plant appears to be one more example of globalization gone unchecked. The factory in Vietnam is one link in a long and mobile supply chain, a link that could be shut down and moved to

China, or Bangladesh, or Guatemala, literally, tomorrow. This factory, which has been operating since the early 1990s, represents the new face of Vietnamese export-oriented industrialization. The factory serves as an export platform, spinning, weaving, knitting, dyeing, and printing imported cotton and polyester fibers into low-value products such as towels for export to developing countries around the region. Part of a wave of Taiwanese, Korean, and Japanese factories at the forefront of the recent boom in foreign manufacturing in Vietnam, factories such as Dona Bochang produce labor-intensive products such as textiles, garments, shoes, and food. These industries have grown at a phenomenal rate, making Vietnam the second fastest growing economy in the world during the 1990s.

Local government officials in Dong Nai (which is the province adjacent to Ho Chi Minh City), have been extremely successful in attracting foreign investment. The province's economy grew by over 30 percent per year during the mid-1990s, providing jobs, tax revenue, and often direct profits to local government officials. And while Dona Bochang clearly produces benefits for some people, the factory also produces serious air and water pollution, and adversely impacts workers' and community health. The dyes used to make Mickey's ears black also turn the local river black (or green, or red, or whatever color dye the factory uses that day). Local groundwater is now too polluted to drink. Air pollution from the factory's boilers regularly coat the surrounding community in a layer of soot.

Despite these problems, for more than six years, local government officials had done little to regulate the factory. From the outside, this case would seem to confirm the worst fears of critics of globalization: local governments have few incentives and even less power to enforce environmental or labor laws that might impinge upon the factory's profits. In interviews, several government officials made clear that they fear firms such as Dona Bochang will move on to other, cheaper countries, if Vietnam regulates them too strictly, or if workers' wages rise too fast. Vietnam, they fear, is already pushing these limits with its $40 per month minimum wage and stricter regulatory controls than other governments in the region.

Community members, however, are less resigned to unregulated industrial activity and pollution. Through letters, meetings, and even protests, community members have repeatedly asked why the plant was allowed to operate in a residential area. The plant is surrounded on two sides by small, tightly packed houses, on the third by the local Catholic church, and on the fourth by an important commercial road, with more houses across the street. Community members on one side of the plant complained that the factory was built with no wastewater treatment system (something that would be illegal for a textile plant in Taiwan, the United States, and most other industrialized countries), and that their wells are now contaminated with toxic wastewater. On the other side of the factory, community members complained about air pollution from the plant, toxic releases in the middle of the night (when no one can see what is being emitted), and the health impacts on their children and older family members.

Few people express surprise at a story of a developing country failing to regulate pollution during its drive to industrialize and connect to the global economy. It is almost taken for granted that governments will pass environmental laws and then fail to enforce them, choosing to postpone environmental protection in the interests of economic development. If this were the conclusion to this case, it would only provide further evidence of the failures of environmental regulation in developing countries and of the inability of local governments to regulate global firms. Fortunately, there is more to the story.

After years of no action, and with clear incentives against enforcement, it has come as a surprise and a relief to the communities around Dona Bochang that the local authorities are now responding to the pollution problems at this factory. In late 1998, after years of community complaints, the environmental agency in Dong Nai required the Taiwanese textile plant to install air pollution control technologies and to change its production practices to prevent pollution. As I explain in detail in the coming chapters, this process involved many starts and stops. Regulating the factory was certainly not easy. But it was—at least with regard to the air pollution issue—ultimately successful. A weak environmental agency effectively negotiated with a powerful and well-connected multinational firm to make changes that cost the firm money.

What explains a successful case such as Dona Bochang under these conditions? Or the case of a Nike factory—perhaps the archetypical example of a global corporation and its footloose suppliers—that has been required through a combination of international pressures and local regulation to significantly improve workplace environmental conditions in Vietnam over the last several years? Or a small, city-owned chemical factory that was relocated against the wishes of local power-brokers after a successful campaign by affected community members?

These cases would be easier to explain if the Vietnamese government was now systematically enforcing environmental regulations throughout the country. But unfortunately, it is not. Like the governments of most developing countries, the Vietnamese government still focuses strongly on promoting industrial development and job growth. In fact, facing new pressures amidst a continent-wide economic crisis, the Vietnamese government has redoubled its efforts to protect state-owned enterprises and promote foreign multinational investments. At the same time, environmental agencies remain understaffed and politically weak.[1] There are no widespread social movements for environmental protection, local NGOs are not truly independent from the government, and there are few market incentives for firms to clean up their production. Many factories continue to operate with little regard for their environmental impacts.

How can successful regulation of industries such as Dona Bochang suggest strategies to advance environmental protection during development? Both successes and the many more common failures of regulation in Vietnam offer useful evidence to analyze when and under what conditions state agencies can enforce environmental laws, particularly under the mounting pressures of globalization. By looking in detail at cases of successful and failed regulation, this research seeks to assess both the potentials for, and obstacles to, regulation of industrial pollution during development. Much of course, is known about the failures of environmental regulation in developing countries; many things can and do go wrong in efforts to protect the environment during development. Cataloguing failures of environmental regulation and examples of governments emasculated by global forces does not, however, tell the full story of regulation around the world. The interesting puzzle motivating this book is that, despite a context that creates strong incentives against regulation and few levers favoring enforcement, cases exist in which state agencies regulate pollution.

This book focuses on this one aspect of environmental protection—the regulation of industrial pollution from manufacturing facilities—in one country. By regulation, I refer throughout to broad social governance of the environmental impacts of industry through state, firm, and civil society actions. Through detailed case studies of successes and failures, we can see when, how, and under what conditions government agencies do and do not regulate. The cases include a range of firms (from a small locally owned chemical factory to a massive Korean factory producing millions of Nike shoes each year), the communities that struggle to survive alongside them, and the government officials in each locale responsible (and struggling themselves) for balancing jobs, development, and the environment.

Vietnam provides an interesting and vexing case through which to examine these issues. Its varying state institutions, wide range of industrial actors, and complicated history of community participation present a window on a number of important theories and debates regarding development and the environment. Because it has experienced such rapid economic development over the last ten years, and has been a literal battleground for development theories over the last half century, Vietnam also brings us face-to-face with pressing debates about globalization, offering a rich context for examining theories of state, market, and community participation in environmental regulation during rapid development.

1.2 Theories of Balancing Development and the Environment

Several broad theories and frameworks are offered to explain environmental regulation during development and to provide strategies for advancing more sustainable development. One major area of research focuses on the role of the state in regulating economic activities through the development of environmental standards and enforcement mechanisms, and points us to the need for building strong environmental agencies with strict sanctioning powers. Another focuses more on market dynamics that can motivate firms to internalize pollution costs and make investments in eco-efficient processes. This framework lends itself to strategies for employing pricing mechanisms to motivate less polluting actions. A third framework stresses the role of public participation in disciplining government agencies and firms.

After thirty years of varied international experiences, however, there is still extensive debate over what processes and dynamics most effectively influence decisions that can control (or prevent) pollution. Questions continue to be raised, for instance, about the effectiveness and efficiency of "command-and-control" (CAC) regulations, and about the equity impli-cations of market-based strategies (Portney 1990, Dwivedi and Vajpeyi 1995, Gottlieb 1995, Desai 1998, Andrews 1999, NAPA 2000). These de-bates center on straightforward arguments about achieving pollution re-ductions at the lowest costs (Cairncross 1992, Opschoor and Turner 1994), to deeper concerns about the constraints and contradictions of policies that try to attract industry, create jobs, and increase tax revenues, while also regulating industry and limiting environmental impacts (O'Connor et al. 1994). Still others have argued that environmental problems are now so complex and changing so fast that traditional state-centric, command-and-control strategies cannot effectively monitor and implement appro-priate regulations (Sabel, Fung, and Karkkainen 1999, Rondinelli 2000). These analyses suggest problems in the design of current policies, and bar-riers to effective implementation and enforcement of regulations (Gibson and Halter 1994).

Although existing theories on policy making (Sabatier 1999) can tell us quite a bit about the processes through which environmental policies are advanced, they do not satisfactorily explain why environmental regula-tions are (and are not) implemented effectively on the ground. Reviewing volumes of postmortems of failed environmental policies around the world and the much rarer success stories provides no clearer picture of exactly when and how state environmental agencies effectively intervene to pro-tect the environment, and what processes influence regulatory breakdown.

This book focuses on the challenge of effective regulation, and in par-ticular on the social and political processes that motivate state, firm, and civil society responses to environment–development problems. It builds on existing theories that focus on individual actors and dynamics—state de-cision makers, the internal workings of the firm, dynamics within social movements—taking particular interest in the critical dynamics that occur in the interstices between actors and across institutional divides.

1.2.1 State Capacity
Environmental theory and policy has historically been state-centric, fo-cused on government programs to protect public goods such as air, water, land resources, and scenic beauty through direct interventions in economic activities. Pollution has been considered an unwanted by-product of legit-

imate activities, and thus the state's role has been to step in to mitigate damage that one set of actors imposes on others, or upon the commons (Weale 1992). States have responded by establishing environmental standards, detailed rules for compliance, and enforcement agencies (Portney 1990, Fiorino 1995). In most countries, standards and environmental agencies are divided along environmental media, with enforcement operating at multiple levels. As Andrews (1992: 2) argues, this has resulted in "a heterogeneous patchwork of diverse statutes, purposes, instruments, agencies, and levels of government."

In a traditional command-and-control regulatory system, the state needs to know what causes the pollution, what levels of emissions are safe, and how to monitor and regulate polluters to these safe levels. Because pollution is a complex phenomenon, however, state agencies have had difficulty gathering and processing the information they need to effectively enforce pollution laws. In an increasingly complex world in which production practices and industrial organizations change continuously, environmental regulators are often unable to keep pace with industrial practices to regulate effectively. Changes in the nature of environmental risks and social responses have led the state to both overregulate and underregulate, focusing too much attention on some issues and not enough on others.

Environmental agencies in developing countries seldom have the technical capability or the political power to enforce compliance with the full range of environmental laws they are assigned to monitor. Low pay and a lack of resources can lead to corruption in the public sector in which fines become a means to supplement incomes rather than reduce pollution. Pervasive corruption not only impedes environmental protection, but also leads the public to lose confidence in the legal system and in the credibility of regulatory agencies, further undermining the rule of law.

A focus on the state as the critical actor leads policy analysts and donor agencies to conclude, quite logically, that what is needed to advance environmental protection is the development of cohesive environmental regulatory agencies with clear policies, sufficient funding, and trained staff, that are insulated from capture by industry and well coordinated with other state agencies. This vision of a sort of "Green Weberian" bureaucracy is common in the reports of policy consultants and environmental donors to developing countries such as Vietnam.[2] Building the capacity of state environmental agencies does, of course, have benefits. Inspectors are better trained, labs are better equipped, and staff members are better compensated (at least during externally funded programs). But building this capacity is extremely difficult, time consuming, and expensive. And the question remains whether building agency capacity alone can lead to regulatory success.

Research in Vietnam would indicate that it is not simply a capacity issue that explains relative enforcement of environmental regulations. The strongest local environmental agency I studied, which received the most external support and had the largest staff, was far from the most active or effective in environmental enforcement. The agency that stood up to Dona Bochang textiles had very few resources, almost no trained staff, and limited political power. Although it is clear that capacity and coherence are important factors in environmental regulation and enforcement, analysis of environmental agencies across countries indicates that capacity alone does not explain variations in environmental enforcement. And as Lester (1995) argues, funding and agency capacity are ultimately the outcomes of deeper social and political processes. The state and in particular local environmental agencies, even when they have training and equipment, are rarely autonomous or powerful enough to implement tough regulations on industry. Put simply, it is not just capacity, but incentives within the state which are critical.

1.2.2 Internal State Conflicts over Environmental Regulation

The preponderance of evidence on regulation in developing countries points to an extremely poor track record for environmental enforcement (Dwivedi and Vajpeyi 1995, Taylor 1995, Desai 1998). A number of commentators have used this evidence to argue that this phenomenon is not just an issue of resources or state capacity. Rather, driven by underlying incentives, states will almost never strictly regulate industrial firms and, even when the state does regulate, it is with significant constraints. Because the state is interested ultimately in the promotion of industry, the accumulation of capital, and taxing these activities (Schnaiberg 1994, Gould, Schnaiberg, and Weinberg 1996), state power and even survival is tied to the promotion and protection of industrial production. The state's interest in promoting industrial activities severely limits its ability to enforce environmental and other regulations that might decrease the profitability of these activities.

Within this framework, policies to promote industry conflict directly with environmental policies. Subsidies to industry, for instance, drive patterns of inefficient resource use and increased pollution that overwhelm antipollution laws. Tax incentives to attract certain industries (such as oil and gas, or chemical production) can create new forms and levels of pollution that environmental agencies are not prepared to regulate. Weak environmental enforcement may actually attract industries fleeing stricter regulations elsewhere. The state may also own and operate production facilities that cause pollution.

Given these conflicts with environmental protection, why do states ever regulate pollution? James O'Connor (1988, 1994, 1998) argues that states are motivated to regulate environmental problems essentially only when the material and social conditions necessary for further capital accumulation are threatened by pollution or resource depletion. In this analysis, the state's main function is the protection of the "conditions of production," and the state intercedes and regulates when these conditions are threatened. Lester (1995) proposes a similar "severity" argument, in which only the most severe pollution problems motivate state regulation.

This argument however, is both too sweeping and too functionalist to explain variations in regulation on the ground. Why for instance, do some state agencies enforce regulations and others not? And why do some firms comply with regulations, while others flaunt them? Processes outside the state certainly influence environmental performance as well.

1.2.3 Market Mechanisms

Over the last twenty years, many analysts have focused on market dynamics to explain variations in environmental performance and regulation. Environmental economists have advocated processes to "internalize externalities" to incorporate the full environmental and social costs of production. This framework focuses both on external market pressures and internal firm responses. Economic theory assumes that firms will respond to external stimuli such as government requirements and market signals, and that firms will take actions to maximize profits within these conditions.

For example, government agencies can motivate firms to reduce pollution or increase resource use efficiencies through policies to "get the prices right." This involves either eliminating subsidies for natural resource use, energy use, and waste disposal, or creating systems of taxes or fees on "environmental bads." A wide range of policies have been implemented in advanced industrialized countries to motivate firms to internalize environmental costs (Stavins and Whitehead 1992, Opschoor and Turner 1994). Even in countries such as Vietnam, where markets for traditional goods have developed only recently, outside advisors and aid agencies have moved quickly to promote market-based strategies to reduce pollution.

Firms obviously respond differently to market signals regarding the environment. However, "rational firms" are expected to reduce pollution whenever it can be demonstrated that changes will be profitable (Prakash 2000). Providing clear market signals and "full-cost accounting" methods are thus hypothesized to support rational decisions to prevent pollution, use resources more efficiently, and to take actions to reduce long-term environmental liabilities. Unfortunately, determining the profitability of environmental investments is often complicated by difficulties of calculating benefits that do not directly accrue to increased sales or decreased costs of production (such as gains in brand reputation or improvements in relations with regulators). And many environmental goods are "public" in nature, and thus difficult for individual firms to control or to justify investing to protect.

Market dynamics are thus only a partial explanation for successful pollution reduction in Vietnam. Regulators in Dong Nai province have not yet implemented pollution taxes, established an emissions trading regime, created an environmental labeling scheme for motivating sales of "green products," or even eliminated subsidies that had supported inefficient uses of water and other resources. As I discuss in detail in chapter 4, market-based mechanisms for environmental protection are still several years away in Vietnam. The market in basic goods and services is still emerging, and while firms are increasingly motivated by market signals, formal market instruments to reduce pollution or shape resource use are still nascent. No regulatory agency in Vietnam has yet attempted real market solutions for pollution problems, suggesting again that other processes are at work in successful cases in Vietnam.

1.2.4 Civil Society Pressures for Environmental Regulation

Another critical influence on state regulatory actions and firm decision making involves civil society pressure. From individual lobbying, to community demands for broader reforms, to full-blown social movements, nonstate actors can influence state policies and their implementation. These pressures can manifest themselves through state-sponsored processes of participation, or through more independent public demands and protests.

States often authorize "public participation." These processes can assume many forms, including formal public comment procedures on Environmental Impact Assessments (EIAs), environmental dispute resolution proceedings, and formal appeals to decision-making bodies. Canter (1996) describes different forms of public participation, ranging from token participation that results in the manipulation of public opinion, to processes where citizens are truly empowered to influence decisions. All too often, official forms of public participation are limited to providing opportunities for the public to comment on development projects through EIAs or other project review procedures. Even in developed countries this input is generally solicited so late in a project's design and development that it rarely leads to changing practices, or to new forms of regulation.

However, some forms of public participation do influence state regulation. Civil-society actors sometimes circumvent official procedures and attempt to pressure firms or agencies directly through protests, boycotts, and informational campaigns (Weale 1992, Fiorino 1995). Chess (2000) describes several such activities in the United States, including public meetings, media campaigns, citizen advisory boards, and citizen juries. In advanced industrial countries such as the United States, the question of civil society influence often boils down to two issues: getting the right participation, and getting the participation right (Chess 2000). Getting the right participation involves getting the right balance of stakeholders at the table for discussions, and assuring that participants are representative of impacted communities. Getting the participation right means assuring that participants are satisfied with the process, that there are generally agreed upon positive outcomes, that the responsible agency responds positively to input, and that the timing of participation is early enough in the controversy to make a difference.

This populist explanation for regulatory variation unfortunately does not fit the cases I examined. In countries such as Vietnam, there is seldom any official public participation in environmental regulatory processes, and no effort to achieve effective or inclusive forms of participation. Vietnam currently does not allow even the most basic forms of participation, such as public review of EIAs. Instead, in Vietnam, and other countries in the region, more spontaneous, unofficial forms of public participation are often responsible for local variations. Struggles to protect local environmental quality by opposing the siting of environmental hazards such as polluting factories, waste dumps, or incinerators are not uncommon. These "NIMBY" ("Not In My Back Yard") battles can evolve into more sophisticated campaigns to broadly oppose environmental hazards and to fight for environmental justice (Gottlieb 1993, Szasz 1994, Taylor 1995), as we have seen in the United States and Europe.

Nongovernmental organizations (NGOs) have also emerged as powerful political and social forces for advancing environmental concerns, as well as for influencing state and corporate decision making (Princen and Finger 1994, Lipschutz 1996, Wapner 1996, Keck and Sikkink 1998). In industrialized countries, environmental NGOs that employ scientists and lawyers have been particularly effective in leveraging state-authorized forms of participation and challenging technocratic policies and programs that previously served to exclude environmental and social concerns. In Vietnam, however, there are few independent local NGOs or large international NGOs working on pollution issues. To my knowledge, no lawsuits have been brought against the government for failure to enforce environmental laws. And there have been no mass action campaigns, consumer boycotts, or national protests regarding the environment.

There is, nonetheless, informal community participation around pollution issues in Vietnam, although these actions look quite different from western conceptions of environmental movements. Under certain circumstances these pressures can play a role in influencing state and firm actions. However, little research has been conducted on how informal public participation influences the implementation of environmental regulations in a country such as Vietnam, or how community participation can influence changes in industrial practices.

1.2.5 Interactions of Actors and Interests

Each of these frameworks for understanding environmental dynamics has clear merits. Community pressures can motivate regulation. Market signals can influence firm decisions. State capacities to command are central to enforcement. However, none of these frameworks on their own is sufficient to explain effective regulation in Vietnam. Implementation of regulations in developing countries is simply "messier" than in idealized government agencies, rational firms, or robust civil societies. In fact, it is the interactions between these actors that are critical to understanding how regulation is actually implemented. Contradictions and competing interests are extremely common in these interactions.

For example, certain agencies and actors within the state have direct incentives to block or delay the enforcement of pollution laws. At the same time, newly established environmental agencies are interested in building professional capacity and power to enforce environmental laws. Jurisdictional issues, coordination problems, and conflicts between agencies responsible for regulation and enforcement can often impede effective regulation.

Firms also have conflicting interests related to the environment. Some firms may seek to externalize environmental costs, polluting as much as they can get away with. Other factory managers may be committed to minimizing production inefficiencies and waste, but face technical, financial, or managerial challenges to reducing pollution. Other firms may view their environmental performance as a competitive advantage (Porter and van der Linde 1995). As has been documented in industrialized countries (Hirschorn and Oldenburg 1991, Fischer and Schot 1993, OTA 1994), many firms are now realizing significant economic gains from pollution prevention and cleaner production strategies.

Community members may also have varied interests and incentives regarding industrial development and pollution. Some community members make their living working in the factories that are the source of local pollution. Others are directly impacted by pollution, losing crops and experiencing illnesses. Still others may actually make their living off of pollution or working in the informal recycling sector. These divided interests in communities make it exceedingly difficult to form collective responses to pollution problems.

Conflicting interests and incentives often lead to situations in which it is easier to accept the status quo of what I call "degrading development," than to advance collective or state strategies for pollution control and prevention. Pollution is simply accepted as a price of development. Countervailing interests and general inertia lead individuals and government officials to choose inaction over environmental measures.

To be fair, no country in the world—either industrialized or developing—has adequately resolved the dilemma of promoting industrialization while protecting the environment. Trade-offs always exist, and economic development often takes priority. And globalization has created myriad new pressures against effective local government regulation.

1.3   Local Regulation in the Global Economy

In an increasingly globalized world, all countries, developing and developed, face external constraints and pressures on their internal economic and regulatory decisions. As countries reach out to connect to international trade and development, they are in turn affected by these new links. Vietnam is very much interested in connecting to the global economy. Kept out of capitalist trading regimes for many years, the Vietnamese government now pins its hopes for development at least partly on increased trade with the world's major economies. The country moved rapidly after the United States lifted its trade embargo in 1994 to establish trade relations with a range of countries, to reestablish lending with the World Bank and International Monetary Fund, and to create the infrastructure necessary to attract foreign direct investment (such as industrial parks, export processing zones, power and telecommunication systems, ports, roads, etc.). Firms like Dona Bochang and Tae Kwang Vina (a Korean Nike producer) have flooded into Vietnam over the last several years. Even domestic producers are now significantly influenced by the global economy as cheaper and higher quality imports are forcing formerly protected domestic markets to open to international competition.

These new connections to the global economy have the potential to both strengthen and weaken Vietnam's ability to regulate pollution. Some analysts predict that countries such as Vietnam will be forced to loosen their environmental standards in order to compete for mobile capital. Others argue that Vietnam's standards may actually be raised by new requirements placed on the government by international treaties, and on producers who seek to sell to international markets.

One set of analysts argues that the decentralization of production across global supply chains in many ways confounds traditional strategies of environmental regulation. Outsourcing production to hundreds of production sites both overwhelms the capacities of local inspectors and brings new pressures against enforcement. Even in industrialized countries, where there are well established regulatory agencies, globalized production systems make regulation more difficult. Furthermore, because these webs cross national borders, efforts to raise environmental standards within one nation may push production to other countries with less stringent standards. This potential for regulatory flight undermines local regulatory powers, and ultimately local sovereignty in a globalizing world. A number of critics of globalization (Korten 1995, Mander and Goldsmith 1996) have argued that corporations are much more powerful in the current context, weakening state agencies vis-à-vis multinational firms, and leaving local movements and campaigns frustrated. Ralph Nader and Lori Wallach (1996: 94) argue further that "corporate globalization . . . establishes supranational limitations on any nation's legal and practical ability to subordinate commercial activity to the nation's goals," and that international agreements such as the World Trade Organization (WTO) effectively establish ceilings on national regulation and "promote downward harmonization of wages, environmental, worker, and health standards" (p. 106).

So are governments completely eviscerated by globalization? Is a "race to the bottom" destined to undermine environmental and other regulations around the world and particularly in countries such as Vietnam?

An alternative reading of globalization asserts that economic integration can benefit and strengthen local regulation. This "trading up" perspective, to use David Vogel's (1995) term, views global economic integration as a tide that can raise all regulatory boats (or at least those that connect into trade regimes). Globalization is not a problem from this perspective, but rather a key to raising environmental, labor, and other social standards. Vogel hypothesizes a "California effect" for environmental regulations where increased trade actually leads states and nations to raise their standards—primarily product standards—to meet the demands of large markets, such as California. So while states are competing for increasingly mobile capital, Vogel (1999), Stewart (1993), and others argue that there is no "race to the bottom" on either product or process environmental standards. If anything, there is a harmonization upwards for environmental standards among key trading partners.

These analysts admit that there are countries that are not being lifted by international trade, countries that they describe as "stuck at the bottom." There are also sectors of domestic production and consumption that never benefit from global standards. Nonetheless, they argue that on the whole, there is an upward ratcheting of environmental standards due to economic integration. From this perspective, further integration into international trade regimes will actually strengthen government efforts to regulate environmental problems.

A third set of analysts also sees benefits of an emerging process of "grassroots globalization" (Karliner 1997). Civil society responses to globalization—as seen on the streets of Seattle, Washington, DC, Prague, and Davos—not only represent a backlash against corporate globalization, but may also be a harbinger of new conflicts over trade and the environment, and new demands from citizens for more regulation of the environmental and social impacts of freely flowing capital. These trends in "counterhegemonic globalization" (Evans 2000) may bring other benefits to communities and the environment around the world, connecting activists and citizens, and identifying local points of leverage over global production.

Vietnam is an excellent country in which to evaluate these different theories of globalization. Have the pressures of joining the global economy forced Vietnam to loosen its enforcement of environmental and other regulations? The answer is probably both yes and no. There are definitely downward pressures on regulation in provinces such as Dong Nai. Foreign multinational managers have made clear that if regulation becomes too strong or labor costs rise too fast, they are more than willing to move their factories elsewhere. But over the last several years, Vietnam has also passed a range of environmental laws and standards, and has likely benefited from increased global awareness of environmental issues since the Rio conference on sustainable development in 1992. The Vietnamese government is aware of the concepts of sustainable development and global environmental protection, understands the environmental problems of neighboring cities such as Bangkok, Taipei, Seoul, and Jakarta, and is under pressure from international donors and institutions to establish strategies that learn from others' failures.

Once again however, this analysis is too broad to tell us much about the success or failure of local regulation. We need to look with a finer resolution lens into Vietnam to see how global dynamics really affect environmental enforcement on the ground, and what kinds of regulations local governments can enforce without spurring capital flight or facing the wrath of local constituents.

1.4 The Research Process

This book analyzes how one country is dealing with the challenges of environmental regulation within a global context, looking seriously from the top-down (state decision makers), to the bottom-up (impacted communities), to the outside-in (transnational activists and international organizations). This research shows that processes do exist for promoting environmental regulation during development, processes that build on state regulation, market signals, and public participation, even in environments where these institutions are not well developed.

Vietnam is an excellent context in which to examine the challenges of balancing development and the environment. The country provides a clear example of an industrializing nation that is seeking at once to promote economic development, environmental protection, and social equity. Although the country was kept out of the global system for many years, between 1990 and 2000 Vietnam moved rapidly to expand international linkages and thus presents an unparalleled case of rapid development and environmental change. It also offers a hard test for evaluating strategies of environmental protection. Successes in Vietnam have occurred where there are no independent local NGOs, no free press, no vulnerable elected officials, and there exists weak environmental agencies, an insulated bureaucracy, and a strong prodevelopment bias.

Yet the Vietnamese government has embarked on a series of initiatives—some externally developed and others internally motivated—that seek to address the challenges of environmental regulation in the current economic development context. Although these have often been frustrated or blocked outright, they have also resulted in some successes. Vietnam is an interesting "natural experiment," a clear case of a country that is newly joining the global economy and making some effort to regulate environmental problems. Vietnam provides a window on both the opportunities and challenges of regulation in the global economy, and varying outcomes in attempts to balance development and the environment.

Ineffective implementation of environmental regulations remains the norm, however, enough successful examples exist under conditions that would predict failure to indicate the potential of processes to promote more effective regulation. This research has sought to analyze these processes and to understand when and how regulation can be effective. Understanding and explaining industrial environmental dynamics requires analysis of multiple issues and actors. The causes of pollution and the motivations for state, firm, and community responses are causally complex. This study employs a research method that Michael Burawoy (1995) calls "reflexive social science." It requires analyzing government policies and processes, the environmental performance of specific firms, different pressures on firms to reduce their pollution, and the roles of community members and nonstate actors. It seeks to engage social controversies, unpack situational experiences, and then to build on existing theories through an analysis of empirical evidence.

The goal of this work is thus to build on and refine existing theories of balancing development and the environment through the case of Vietnam. The research examined actual cases of environmental regulation and developed hypotheses to explain when and how this regulation is effective. Based on this research, I propose a tentative hypothesis or "model" of effective regulation, which I call Community-Driven Regulation.

The Community-Driven Regulation (CDR) hypothesis asserts that state command-and-control regulation and market dynamics are critical, but not sufficient to explain variations in environmental regulation in Vietnam. CDR stresses the interactions between state actions, firm dynamics, and community demands. The cases I present show that when communities see, smell, or feel pollution and then mobilize to effectively pressure a state agency, it is possible to advance environmental regulation that balances development and the environment. This is not to say that community actions alone determine the effectiveness of regulation. Rather, it is community pressures that kick in motion a dynamic between agencies, NGOs, and firms, that ultimately determine the effectiveness of regulation. I stress the role of community participation in this process because it has been largely overlooked in the literature on environmental protection in developing countries. There are of course a number of contingent elements in these processes that require further theorization and study, but the preponderance of evidence from my research points to this dynamic of Community-Driven Regulation.

This research is, however, preliminary in a number of regards. The analysis is focused on a limited range of problems and dynamics in one country. The analysis—and thus the relevance of Community-Driven Regulation—is limited to pollution problems that are clearly perceptible and relatively concentrated. Advancing the CDR hypothesis is meant to further discussions on environmental politics in developing countries rather than to conclude them.

1.4.1 Methods

The findings of this research are based largely on a comparative analysis of six case studies in Vietnam. Each case was centered on a factory and its pollution—analyzing the communities impacted by the pollution, the state agencies responsible for promoting and regulating the factory, and nonstate actors who had a role in influencing the factory's environmental performance. Cases were selected from an initial pool of over 120 factories.[3] The six cases chosen for in-depth analysis in this book were selected to evaluate different explanatory variables, such as: ownership (foreign versus Vietnamese managers); market dynamics (such as whether the company competed internationally or domestically); the strength of the local regulatory agency; the cohesion of the neighboring community; and, types and levels of pollution impacts (e.g., air versus water pollution). Cases were drawn from both the north and the south of Vietnam, which have had historically very different political systems, different forms of community organizations, and different levels of private ownership of industry.

Case selection was carefully designed to avoid the hazard of "selecting on the dependent variable." Cases were not selected to "prove" the CDR hypothesis. Rather, the initial research design focused on identifying and studying different types of firms and state agencies. Two foreign-owned firms, two centrally controlled state enterprises, and two locally managed firms were selected. And three different jurisdictions (Dong Nai, Phu Tho, and Hanoi) were selected to evaluate variations in state action. I was, of course, aware of and interested in exploring variation in the dependent variable—effectiveness of regulation. Without some successes and some failures, it would have been extremely difficult to tease out the factors that influence either. Table 1.1 summarizes the basic features of the firms in the case studies.

The background for each of the six case studies included analysis of local institutions and politics, relevant Vietnamese history, national development programs, and environmental policies. An early stage of the research also involved a detailed review of relevant laws and institutions, and an analysis of international economic changes and pressures affecting industrial development and environmental decision making in Vietnam. The research was conducted and analyzed within a broader concern for global connections and pressures.

The research was designed to be explicitly multidisciplinary, combining technical, policy, and sociological analyses. Waste audits and technical assessments of pollution sources and impacts were conducted. For each case, I gathered information on environmental trends, company characteristics and actions, community characteristics and actions, state and local regulatory roles, and market dynamics. A comparative institutional approach was employed to analyze similarities and differences among the cases.

1.5 Central Themes and Dynamics

The encouraging finding of this research is that it is possible to break a cycle of degrading development, and to begin to work towards balancing industrial development with environmental protection. The qualification is that this does not happen very often, and is only successful under specific circumstances. But when the conditions exist (or can be created) and community actors mobilize to pressure state environmental agencies to regulate the environmental impacts of industrial activities, a different form of development, perhaps even the seeds of sustainable development, appears to be possible.

Table 1.1
Overview of Case-Study Firms

	Dona Bochang	Lam Thao	Viet Tri	Tan Mai	Ba Nhat	Tae Kwang
Product	Textiles	Fertilizer and chemicals	Chemicals	Paper	Chemicals	Shoes
Output per year	18 million towels	570,000 tons Superphosphate 180,000 tons Sulfuric Acid 3000 tons insecticide	5,000 tons Chlorine gas 6,000 tons NaOH 8000 tons detergent	68,000 tons paper	3600 tons Calcium carbonate	6 million pairs Nike shoes
Employees	900	3000	650	750	200	9,200
Ownership	Taiwanese-Vietnamese joint venture	Ministry of Industry	Ministry of Industry	Ministry of Industry	Hanoi Department of Industry	Korean
Pollution	Boiler gases Soot Dyes	Acids Sulfur dioxide Sulfuric acid	Chlorine gas Sodium hydroxide detergents	Black liquor Boiler gases Fibers, dust Chemical oxygen demand	Calcium Carbonate dust Noise	Boiler gases Solvents Solid waste
Location	Dong Nai province New urban	Phu Tho province Rural	Phu Tho province Semi-urban	Dong Nai province New urban	Ha Noi Urban	Dong Nai province Industrial estate
Established	1990	1962	1961	1963	1968	1995

Through Community-Driven Regulation, community actions play a central role in pressuring state environmental agencies to improve monitoring and enforcement capabilities, prioritize environmental issues for regulatory action, directly pressure firms to reduce pollution, and raise public and elite awareness of environmental issues and the trade-offs between development and the environment. Examples of Community-Driven Regulation display a dynamic of mutually reinforcing interactions between organized communities and state environmental agencies, pointing to the potential for "state–society synergy" (Evans 1996) and to a transition from traditional "command-and-control" regulation to a more responsive system of "demand-and-control" regulation.

In this process of community demands and government controls, interactions between communities and state agencies are obviously critical. The process of creating a dynamic of coproduction of regulation (Ostrom 1996), and sequencing interactions between communities, state agencies, and firms are central to making regulation work in Vietnam. Capacities of state agencies matter in these processes. But so do community cohesiveness and mobilization, linkages to outside actors, and means of interaction between the state and civil society.

Politics within the state also matter. The state is not monolithic when it comes to environmental issues. The relative power of one agency over another can influence whether regulation succeeds or fails. Community pressures and other outside forces can serve to tip the balance in these internal political battles. The presence and sequencing of pressures is also critical as some pressures have more powerful triggering effects than others, with some leading to effective regulation against existing forces.

In the Dona Bochang case and several others described, it was not simply an improved capacity of the state agencies, or a change in a market signal, or even a successful public campaign, but rather the interaction between communities and the state that finally tipped the balance towards enforcement. Through a long and complicated process, a certain synergy was achieved between the local state agency and the communities that motivated environmental regulation. This complex dynamic is at the heart of what is analyzed in the following chapters. It involves conflict, cooperation, and sometimes coproduction of environmental protection. It can be derailed at many turns, but it can also, under certain conditions, achieve the goal of promoting more ecologically sustainable development.

The process of Community-Driven Regulation can also support and strengthen state institutions for environmental protection. For example, the community members around Dona Bochang have not only successfully pressured the firm and the government to reduce the air pollution from the factory, they have also played an important role in strengthening the local government agency's ability to regulate industry more broadly. Community demands and pressures have forced the local government to improve its capacity to monitor and enforce environmental laws. In another case I discuss in detail, community activists in the United States who have been pressuring Nike and its global production network are in different but similarly important ways influencing firm performance and government actions in Vietnam. Both sets of actors work to chart a course that more effectively balances industrial development with environmental and social concerns.

The actions of community members in Vietnam, and communities of concerned citizens in other countries, are by no means always successful. In fact, many community attempts at environmental protection fail. Nonetheless, my research identifies a pattern of success that deserves further examination. Much can be learned from existing processes—both successes and failures of regulation.

The empirical evidence presented in the coming chapters directly engages a number of debates about environmental regulation and the challenges of promoting "sustainable" industrial development. Much of the existing literature ignores the underlying social and political processes that lead to failures of environmental regulation, and that can motivate state and civil society actions to effectively control or prevent degrading development. By examining interactions between state agencies, firms, and communities involving specific cases of industrial pollution in Vietnam, this book presents an empirically grounded analysis of the politics of environmental regulation and of balancing development and the environment.

The Community-Driven Regulation model builds on traditional understandings of state environmental regulation. It shows how community participation is critical to motivating the state to act, and how public participation can actually strengthen weak environmental agencies. The cases also show how market dynamics, and motivations for firms to take environmental actions, are often undergirded by thick institutional and political processes.

This research also directly addresses pressing policy questions. The cases in Vietnam speak to challenges of minimizing adverse impacts of industrialization, and to questions about mechanisms for promoting pollution prevention during development. The study looks beyond traditional regulatory strategies to analyze innovations from different sectors and to learn from successful processes where they exist.

In many countries, policy analysts fall back on the easy answer of trying to build institutions similar to those that exist in the industrialized world, for instance, creating U.S.-style Environmental Protection Agencies in developing countries. Unfortunately, international efforts to strengthen environmental regulation in countries such as Vietnam often fail to analyze and address the underlying motivations for, and impediments to, environmental enforcement. Most international aid remains focused on passing more and better environmental laws and building the capacity of state environmental agencies in order to strengthen "command-and-control" institutions, without understanding the political and economic dynamics driving decision making on the ground. Similarly, many international projects and aid programs promote market-based instruments as a cure-all for past failures of environmental regulation without understanding the politics underneath market processes in these countries. These forms of international assistance are often doomed to frustration or failure.

The case studies presented in this book offer examples of successful regulation in an extremely challenging context. Understanding how these work, and then building on them, seems a more promising strategy for future regulation in Vietnam (and other countries) than attempting to import models from abroad. Processes such as Community-Driven Regulation have the potential to be even more effective if they are supported by intentional policies that foster the underlying mechanisms of community participation and state responsiveness. My hope is that by laying out how Community-Driven Regulation is currently working in Vietnam and examining its key actors, interactions, and dynamics, this book will add to debates about advancing more ecologically sound and socially just development.

1.6 Roadmap for the Book

The goals of this book are ambitious: to describe in detail the specific processes at work in community, state, and firm responses to pollution issues in Vietnam and to build on existing theories and hypotheses about how these dynamics can serve to reduce pollution during industrial development. I do not attempt to statistically prove causality in these processes, but rather explore the complex dynamics around industrial pollution problems in enough detail to draw some conclusions about how things work on the ground and what factors appear to be critical to successful pollution reduction.

Rather than structuring the book around the six case studies, which might seem logical, I have chosen instead to organize the argument around the key actors involved in the processes. My goal is to examine in detail the key institutions at work in existing conflicts over industrial development and pollution, and to examine the conditions and contexts necessary for successful environmental regulation. The focus on actors facilitates an analysis of institutional processes—uncovering underlying motivations and incentives for action, and conflicts within and among different actors. This framework also helps elucidate the processes by which state agencies and firms can be motivated to take action on pollution issues, and the role that community members and other actors play in promoting more ecologically sustainable industrial development.

Chapter 2 provides an introduction to the critical challenges of balancing economic development and environmental protection in Vietnam. The chapter discusses the extremely rapid economic changes underway in Vietnam—establishing the country as the next "Tiger" economy of Asia—the mounting environmental problems the country faces, and existing regulatory responses. The chapter concludes with a brief discussion of successful examples of environmental regulation within the very challenging Vietnamese context.

Chapter 3, "Cohesion, Connections, and Community Actions," discusses the critical roles community members play in Community-Driven

Regulation. The chapter discusses theories of social mobilization, under-standings of community in Vietnam, and the origins and effectiveness of community mobilization around pollution incidents. The chapter draws conclusions about the social, cultural, and historical bases of effective community action. Issues of internal community cohesion and external linkages emerge as central to the success of the cases.

Chapter 4, "Plans, Profits, and Pollution Decisions" takes us inside the firms to examine the decision-making processes that ultimately determine whether pollution is reduced or not. The chapter discusses a range of char-acteristics and responses of firms and some conclusions are drawn about which types of firms are most susceptible to community and state pres-sures, and how communities and regulators can be most effective in moti-vating firm decisions.

Chapter 5, "Motivating a Conflicted Environmental State," presents an analysis of the roles and responses of state agencies. This chapter begins with a discussion of theories of state environmental enforcement and then analyzes actual conditions inside frontline environmental agencies to understand their motivations and constraints for regulatory action. The analysis ranges from discussions of broad state legitimacy, to agency re-sponsiveness to community pressures, to basic capacity issues that may hinder or support regulation.

Chapter 6, "Information, Accountability, and Direct Pressure Politics," discusses the impacts of extralocal, nonstate actors in each case. The chap-ter focuses in some detail on the case of Tae Kwang Vina, a Korean sub-contractor for the Nike shoe company, and the dynamic and evolving roles played by transnational advocacy networks spanning from Vietnam to the United States and back. The chapter also discusses the important role the media plays in amplifying the complaints and demands of community members.

Chapter 7 explicates the Community-Driven Regulation model in more general terms, and includes the roles and characteristics of key actors. The goal of the chapter is to present an hypothesis to explain the general processes operating across the cases. Community-Driven Regulation is presented as an exploratory framework for explaining (and possibly ad-vancing) more effective environmental regulation.

Finally, chapter 8, "Regulation against the Odds" concludes with reflections on how Community-Driven Regulation currently functions in Vietnam and what potential exists to further the underlying mechanisms of CDR. The chapter explores the theoretical implications of Community-Driven Regulation and discusses policies and programs that might improve existing processes. The chapter then discusses a range of policy options, some of which are already working in other countries, that might advance environmental protection by strengthening both state and community actions. It places the Vietnamese case in comparative perspective through a discussion of similar policy dynamics and experiments underway in countries such as Indonesia, China, and the United States, and concludes by reviewing strategies to build on the dynamics of Community-Driven Regulation to create greater synergy between state, firm, and civil society actors for environmental protection.

2

Development and the Environment in Vietnam

America is coming to see Vietnam . . . as a country, not a war.
President Bill Clinton, November 17, 2000.

As one of the fastest developing economies in the world and slated as "Asia's next Tiger economy" (Mydans 1996), Vietnam is a critical case for evaluating the challenges of balancing development and the environment. In its transition to a market economy and opening to global trade, the Vietnamese government is assuming new roles in promoting economic development, and at the same time, new strategies for protecting the environment and workers. Promotion of industry is being balanced (often precariously) with the regulation of foreign multinationals, the establishment of new laws and institutions, and continued state control over most areas of politics.

Vietnam offers a sort of "natural experiment" through which to analyze development changes and environmental impacts, and to compare theory to practice. Because the country was kept out of the global economy for so many years—due to wars and the United States trade embargo—it was not until the early 1990s that Vietnam began to take off economically. In a relatively short period, essentially from 1989 to 2000, Vietnam has experienced unparalleled development (based largely on a massive inflow of foreign investment), and environmental change. Vietnam's current development strategy—focusing on labor-intensive export industries and natural resource extraction—and limited regulatory capacity, would certainly lead to predictions of lax environmental enforcement, the acceptance of a "grow now, clean later" attitude, and a low level of public awareness and action around environmental issues.

Vietnam's rapid growth makes it a potentially archetypal example of degrading development. Economic growth remains paramount. The country continues to depend on cheap labor to attract foreign investment. Vietnam's minimum wage of $35 to $45 per month (depending on the province), competes with nations such as Indonesia, Bangladesh, and China. The government continues to promote the exploitation of natural resources and the development of polluting industries to jump-start the economy. At the same time, government agencies responsible for environmental protection are understaffed and politically weak. And government corruption, in its many forms of abuse of public roles and resources for private benefit, remains severe. Vietnam was ranked the fifteenth most corrupt government in the world in 2001 by Transparency International, trailing only Indonesia and Bangladesh in Asia.

Vietnam is, of course, not alone in this predicament. Many developing countries are similarly seeking to rapidly industrialize while also struggling to protect their environments and human health. Countries as diverse as China, Mexico, and Brazil all face similar challenges and trade-offs.

Vietnam, however, is an interesting case partly because of the strong pressures the government faces for balancing industrialization with social and environmental concerns. The country has a long tradition of bottom-up pressures that interact in complex ways with the top-down character of the government. In both economic and environmental spheres, community action has played a driving role in state policies and programs. Because of this historical legacy and current politics, the government does not have the luxury to ignore environmental issues.

From the outside, Vietnam may appear to represent a unique category of nations. It is one of the few remaining communist states in the world. However, as I describe, one decade after a major renovation process called *Doi Moi*[1] was initiated, Vietnam is now in many ways more like Thailand or China, than Russia or Hungary. Vietnam is clearly a "transition" economy. But it is also a developing country, like many others around the world, seeking to make the transition from agriculture to industry to global trade. Experiences on the ground in Vietnam thus have relevance to a wide range of developing nations.

Vietnam is also similar to other developing countries in that one would not expect successful community or state responses to environmental problems. Neither academic or media observers would expect communities in Vietnam to have power, or the state to have capacity and interest to regulate industry. Nonetheless, the state does have some capacity and communities do have some potential to precipitate action.

The purpose of this chapter is to provide some context on Vietnam, and more specifically on the struggles between economic development and environmental protection currently at play. Vietnam is in some respects a "success story" of development, and an emerging "horror story" of environmental problems. This chapter provides background on these trends and recent government responses for better protecting the environment during development. Understanding this contested terrain, and how regulation can sometimes work within this context, provides insights into the potential for more effective environmental regulation both in Vietnam and in countries with greater endowments of state capacity or social capital.

2.1 The Vietnamese Context

Just to provide the most basic context: Vietnam is poised on the edge of Southeast Asia, with China to the north, and Laos, Cambodia, and Thailand to the west. As the map in Figure 2.1 shows, Vietnam is endowed with a coastline that stretches over 1,500 kilometers. The country is approximately 330,000 square kilometers in size, or about three-fourths as large as California. However, recent estimates put Vietnam's population at just over 80 million people (2.5 times that of California), making it the thirteenth most populous country in the world.

While population growth has slowed considerably—from 3.1 percent per year in the 1970s to 1.8 percent in 1997—the working age population continues to expand. An estimated 1 million adults are added to the workforce each year (EIU 2000a). Creating jobs for these people is a top priority for the government. Although the official urban unemployment rate was 7 percent in 1998, an additional 15 percent of the population is estimated to be underemployed (EIU 1998).

Vietnam's annual per capita GDP is estimated to be around $300 (EIU 2000a), which is more than double the GDP recorded in 1993. However, income disparities between urban and rural areas are growing. Annual per capita income in urban areas was estimated to be approximately $650, with Ho Chi Minh City—the richest city in the country—reaching per capita GDP of $1,365 in 2000 (DPI 2002). Per capita GDP in rural areas

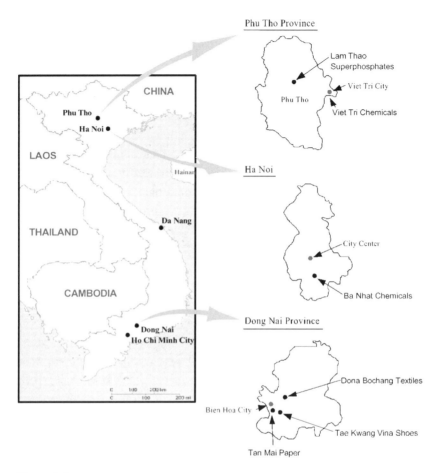

Figure 2.1
Map of Vietnam and case study provinces

however, remains only around $180. And a large percentage of the population earns its income in the informal sector.

2.2 Economic Development in Vietnam

As recently as the early 1990s, Vietnam was one of the poorest countries in the world. This is changing; however, despite the fact that the populace remains overwhelmingly rural, with over 75 percent of the population living in rural areas. Much of this rural population is concentrated in two rice growing areas: the Red River Delta (in the north) and the Mekong Delta (in the south). A postwar population explosion has made these areas, and Vietnam overall, one of the most densely populated "rural" countries in the world, as figure 2.2 highlights.

Vietnam's economy has experienced several major transitions over the last fifty years. In a number of regards, Vietnam has been a literal battleground for development theory. From the French to the Russians to the Chinese to the Americans, and more recently to institutions such as the World Bank, Asian Development Bank (ADB), and International Monetary Fund (IMF), Vietnam has been "assisted" by competing theories and programs of development.

The most recent round of development changes began at the conclusion of the American war and the reunification of the country in 1975, when the government of the Democratic Republic of North Vietnam extended its economic model to the south. This involved the imposition of central planning, collective agriculture, land reforms, state-ownership of industry, and controlled trade with the Soviet bloc—the COMECON countries.

Figure 2.2
Plowing a rice field

This economic model was almost immediately problematic. During the 1980s agricultural productivity declined, price and wage disparities created widespread discontent among the citizenry, inflation surpassed 300 percent, industrial output failed to meet targets, and the country defaulted on IMF loans (World Bank 1993). External pressures on Vietnam's economy were mounting as well. Major changes in economic policies in the Soviet Union, Eastern Europe, and China during the 1980s included new pressures for Vietnam to be more productive and competitive. These changes however, also opened up ideological room for changes in Marxist–Leninist economic theory in Vietnam (Ljunggren 1993). Changes in Soviet aid programs, and in particular pressure to more effectively use Soviet finances and technology, also forced the Vietnamese government to rethink the structure of its economic system. The growing trade imbalance with the COMECON was a further impetus for the government to focus attention on export-oriented manufacturing away from self-reliant heavy industry (SarDesai 1992, Fforde and de Vylder 1996).

A political and economic "renovation" was thus initiated in 1986 and accelerated in 1988 after a famine shocked the country. This process, known as *Doi Moi,* set out to achieve three primary goals: reduce dependence on the COMECON; improve macroeconomic management of the economy (including controlling inflation and increasing productivity); and attract foreign investment. The immediate desire to satisfy basic needs for food staples was also a critical priority. The government accomplished this by creating conditions for private agricultural production, and by allowing peasants to sell surplus production to the open market. Price controls for most goods were lifted in 1989. By the early 1990s the government was moving swiftly to support the marketization, privatization, and internationalization of the economy (Fforde and de Vylder 1996).

In a very short time span, Vietnam thus moved from autarkic self-sufficiency with limited trading, to export-oriented production as its focus of economic development. The country has also turned outwards to foreign capital and western aid for financing and technology to develop its industrial base. Hundreds of state enterprises have been shut down, and others are being encouraged (or forced) to compete in international markets. Changes in the agricultural sector were so successful that Vietnam moved from being a rice importer to being the number-three rice exporter in the world in a matter of years. Inflation fell from 300 percent to 35 percent in one year. Government revenues rose from newfound taxes and expanding oil exports. A 1990 World Bank report called Vietnam's reforms "the most comprehensive and radical set of reform measures adopted by any socialist country at the time" (Ljunggren 1993).

On top of the *Doi Moi* reforms, the Vietnamese government has advanced a program of "Modernization and Industrialization" designed to respond to the agricultural and employment crisis in the country. This philosophy of development has been supported by a wide range of policies, laws, and directives, and has led to major transformations in the country's economic, social, and environmental systems. Since the lifting of the U.S. trade embargo in 1994, and the subsequent normalization of World Bank and IMF lending, Vietnam's economy has begun to resemble the Tiger it aspires to be, albeit with some fits and starts.

While Vietnam as a whole has experienced GDP growth of over 8 percent per year throughout the 1990s, with industry growing by 13 percent per year, Vietnam's urban centers have grown at twice that rate (EIU 1997 and 1998, 2000a). Even amidst the Asian economic crisis, when overall GDP growth slowed to approximately 4.8 percent in 1999, Vietnamese industry continued to expand by over 10 percent (EIU 2000b). In 2000, the economy regained momentum with overall GDP growing by 6.8 percent and industry growing by a whopping 15.5 percent. Table 2.1 shows overall GDP and industrial growth figures from 1993 through 2001 for Vietnam. To give some perspective, in 1999, Indonesia's economy actually shrunk by 13.7 percent while Thailand's economy contracted by 8.0 percent.

A massive inflow of foreign direct investment (which has slowed since 1998) and foreign aid (which has stayed almost constant at $2 billion per year from 1994 to 2000) has been driving a process of rapid urbanization and industrialization. The Vietnamese economy however, is not just growing rapidly, it is also being transformed. Economic reforms and foreign investment have led to a significant shift in the structure of the economy, and a troubling trend toward more toxic industrial activities. Small- and medium-sized enterprises (SMEs) are multiplying in cities around Vietnam, while large foreign joint ventures are concentrating in the country's more than fifty new industrial zones. The oil and gas, steel, chemicals, garments, footwear, and printing sectors have all grown by over 20 percent

Table 2.1
GDP Growth (%)

	GDP growth (%)	Industrial Growth (%)
1993	8.1	13.2
1994	8.8	14.0
1995	9.5	13.9
1996	9.3	14.4
1997	8.8	13.2
1998	5.8	11.0
1999	4.8	10.4
2000	6.8	15.5
2001	6.5	10.5

Source: Asia Pulse (2001), EIU (2001)

per year during the 1990s. Many of these industries are, not incidentally, serious polluters.

There has also been a shift in the management of industry. In the past, almost all major industries were controlled by either central or local state authorities. Today, the private domestic and foreign invested sectors are the fastest growing segments of the economy. In 2000, the emerging private sector recorded a growth rate of 18.6 percent, with foreign firms trailing slightly with 17.0 percent growth (EIU 2000b).

Rapid urbanization is occurring simultaneously with industrialization. Vietnam's urban population is currently estimated at 15 million people, or 20 percent of the total population. However, urban population growth has been estimated at 4 percent per year, almost triple the rural population growth rate. Approximately half of the urban population is concentrated in and around Hanoi and Ho Chi Minh City. Past policies to control urban growth, such as restrictions on migration, planned decentralization of industry, and the development of "New Economic Zones" in the periphery, are becoming less effective or being abandoned altogether.

The government now actually appears to support, or at least accept, further urbanization. As figure 2.3 symbolizes, this has led to a rapid integration of old and new, rural and urban throughout Vietnam. According to the Ministry of Planning and Investment (MPI), the government plans for 40 percent of the country (an estimated 37 million people) to be living

Figure 2.3
The collision of old and new on the streets of Ho Chi Minh City

in cities by 2010. Growth on this scale and pace in urban areas has the potential for significant social and environmental problems. The government's primary strategy to avoid this problem, and more specifically to avoid the creation of "mega-cities" such as Bangkok, Jakarta, and Shanghai, is to evenly distribute development throughout Vietnam. Three growth triangles are being advanced in: (1) Hanoi-Haiphong-Quang Ninh in the north; (2) Quang Nam-Danang-Dung Quat in central Vietnam; and (3) Ho Chi Minh City-Dong Nai-Vung Tau in the south. Regional gaps in economic development remain a major concern for the government.

Central government officials plan urbanization in coordination with further industrialization, meaning a reduction in agricultural employment and an increase in manufacturing and service sectors. As an indication, 100 industrial zones are slated for operation in the next ten years. Table 2.2 shows the planned shift in the structure of the Vietnamese economy by the year 2010.

The government recently published new economic targets that call for GDP growth of 6 percent per year from 2001 to 2005, then increasing to

Table 2.2
Planned Changes in Structure of the Economy (% GDP)

	1996 (GDP breakdown)	2010 (GDP breakdown)
Agriculture	26.2%	17%
Industry	31.3%	37%
Service	42.5%	45%

Sources: EIU (1998) and Ministry of Planning and Investment Development Strategy Center (1999).

7 percent per year from 2006 to 2010 (EIU 2000a). Industrial growth is slated to accelerate from 7.8 percent for 2001 to 2005 to 9 percent per year from 2006–2010.

The signing of a bilateral trade agreement with the United States in July 2000 symbolized another major milestone in Vietnam's economic transformation. This agreement, which many argue is a prerequisite to membership in the WTO, will involve further liberalization of the Vietnamese economy, and further economic growth. The main benefit to Vietnam will be the lowering of tariffs on most Vietnamese exports to the United States from 40 percent to approximately 3 percent. The World Bank estimates this arrangement will boost Vietnam's exports by up to $800 million per year.

These changes in the economy have had clear benefits. Life expectancy has risen from forty-three years in 1960 to sixty-eight years in 1994. Child malnutrition has decreased drastically from 53 percent of children as recently as 1993, to 34 percent of children in 1998. The number of people living in poverty has also declined from 23 percent of the population in 1993 to 15 percent in 1998 (EIU 2000a). At the same time, while the economy as a whole—and some groups in particular—are benefiting from economic developments, inequality is also on the rise in Vietnam (EIU 2000a).

2.3 Impacts on the Environment

Vietnam's phenomenal economic development over the last ten years, while obviously positive in many regards, has had major impacts on the country's environment. Urban and industrial development activities have resulted in a wide range of impacts on land use patterns, resource depletion, biodiversity loss, and water, air, and soil pollution. At the same time, environmental conditions and natural resources have, in turn, influenced the country's development processes, such as human migration and industrial development patterns. In the long term, interactions between industry and the environment may ultimately boil down to a question of whether environmental degradation and resource depletion actually undercut Vietnam's development strategy.

Urbanization processes are changing environmental and resource conditions in both urban and rural areas. Land use is being altered in major cities and on their peripheries as rural land is converted to residential, business, and industrial uses. Hanoi and Ho Chi Minh City continue to expand into formerly rural areas. Concern about rapid land use changes led the government recently to pass a decree prohibiting further conversion of farm land to industrial uses.

Rapid growth in industry is creating new stresses on uses of natural resources and the environment. Growth of 10 to 14 percent per year in industrial activities (with some sectors more than doubling in the last few years) is requiring increased extraction of natural resources, increased production and use of energy, and more transportation and other infrastructure services, all of which result in more waste and pollution. Changes in the scale and structure of the economy, the efficiency of industrial activities, and mechanisms of regulatory control, all affect rates of natural resource depletion and pollution levels. Increased demand for energy also leads to a wide range of impacts on land uses, natural resource extraction, and air and water pollution.

Natural resources are clearly being impacted by these economic developments. Forest cover has declined from 44 percent of the country in 1943 to less than 20 percent today (Poffenberger 1998). Over 1.5 percent of forest land is being deforested each year, a rate that will eliminate all remaining forests within fifty years (EIU 2000a). Marine ecosystems are similarly being damaged and depleted by overfishing, while mangrove forests are being threatened by new shrimp farms.

Vietnam is also experiencing a shift in the structure of industry towards more polluting sectors, and from traditional pollutants [such as biochemical oxygen demand (BOD)] to complex toxic compounds (such as heavy metals and hazardous wastes). Extractive industries such as oil and gas production, and mining, which have been expanding rapidly, cause significant environmental impacts. Other resource-based industries such as food processing and aquaculture also result in increased pollution loads to rivers. In the future it is likely that highly polluting sectors such as petrochemical production will bring new hazards to Vietnam. The World Bank has estimated that "if Vietnam does not implement pollution prevention and control policies, its toxic intensity will increase by a factor of 3.8 over a ten-year period (2000–2010), equivalent to a 14.2 percent annual [pollution] growth rate" (World Bank 1997).

There are currently approximately 300 medium- and large-sized factories operating in Hanoi and 700 medium- and large-sized factories in Ho Chi Minh City. Over 90 percent of these factories have no waste treatment facilities. Based on an analysis of factories in Hanoi, the city's Environment Committee reported that 124 of the 300 factories had emissions exceeding environmental standards (UNDP 1999). Of sixty industrial zones or export processing zones in the country, only three have dedicated wastewater treatment plants. And even as Hanoi and Ho Chi Minh City begin to strengthen their environmental regulations, polluting industries are expanding to rural provinces. One newspaper account recently reported that "while toxic industries have been rejected in certain provinces, they have been welcomed by others with easier procedures" (Asia Times 2000).

Water pollution is a growing problem throughout Vietnam. Many rivers and canals near urban centers are burdened with municipal and industrial wastes. Water emissions from sectors such as food processing, beverages, textiles, paper, and chemicals include: biochemical oxygen demand (BOD), chemical oxygen demand (COD), acids, chlorinated organics, and heavy metals. Groundwater in cities such as Hanoi is also contaminated. Virtually all of the domestic wastewater in Vietnam is discharged untreated into rivers. Industrial effluents are also often still discharged without proper treatment, affecting both ecosystems and human health. Water pollution from industry regularly leads to fish kills, crop damage, and a wide range of skin and stomach illnesses in communities living near industrial zones. A recent report by a Canadian consulting firm found concentrations of certain toxic chemicals (including pesticides) at 890,000 times the allowable levels for California drinking water standards (Hatfield Consultants 1998).

Air pollution in urban areas is worsening with increased manufacturing activities, energy production (particularly from coal), and vehicle emissions. Dust pollution (particulate matter) has reached alarming levels in many urban areas due to construction projects, increased cars, trucks and motorcycles, and industrial activities. Toxic air pollutants have been measured at unhealthy concentrations near industrial facilities located in residential areas. A Ministry of Science, Technology, and Environment report in 2000 reported over 4,000 industrial entities were releasing smoke, dust, and toxic gases above legal limits.

Solid and hazardous wastes are also on the rise with the growth in household and industrial consumption. Sectors such as steel, electronics, and chemical manufacturing are producing new sludges, acids, and toxic solvent wastes. The government estimates that industrial estates alone produce approximately 400 tons of solid waste per day. Vietnam as a whole generates approximate 7 million tons of industrial waste per year (Interpress Service 2001). Unfortunately, there are no systems in place in Vietnam for handling, storage, or treatment of hazardous wastes. An alarming amount of hazardous wastes from industry and hospitals are thus released untreated (or illegally dumped) in urban areas (JICA 1998).

Volumes of residential wastes are also increasing rapidly. Consumption habits are beginning to create the outlines of a Vietnamese throwaway society, while domestic solid waste collection and disposal remain grossly inadequate. Vietnam currently generates approximately 8 million tons of solid waste per year. Most estimates are that only 50 percent of municipal solid waste is collected and disposed of properly. Landfills that do exist do not meet international standards for safety and management. And as figure 2.4 shows, informal "recycling" or scavenging is common at waste dumps.

Workplace environmental issues present severe problems in industries throughout Vietnam as well. Workers are commonly exposed to toxic air pollutants, noise, heat, and radiation, without proper protections (Ministry of Health 1998). An estimated 20,000 workers in Vietnam suffer from silicosis, and over 10 percent of all industrial workers are estimated to be exposed to harmful noise levels (MoH 1998). Four hundred and three workers were reported killed in industrial accidents in 2000, although the government admits the real figure may be ten times higher (Deutsche Presse-Agentur 2001). Use of chemical fertilizers and pesticides has skyrocketed

Figure 2.4
Scavenging for materials in a trash dump near Haiphong fertilizers

over the last decade, with chemical fertilizer use (per cropped hectare) increasing from 40 kg in 1982 to 223 kg in 1997 (EIU 2000a). Pesticide poisoning is growing among farm laborers as Vietnam continues to permit the spraying of toxic pesticides such as hexachlorobenzene, Dieldrin, and DDT. These pesticides, banned in most developing countries, also adversely affect groundwater and soil quality.

Energy production and use is an important contributor to pollution and resource depletion. Vietnam currently has one of the lowest levels of energy and electricity consumption in the world at 130 kilograms of oil equivalent and 350 kilowatt hours (kwh) per person (about one-fifth of Thailand's consumption). However, both production and consumption of energy are increasing rapidly (England and Kammen 1993, EIU 2000a). Researchers in Hanoi estimated that energy consumption will grow at a rate of 10 percent per year in the north, and 15 percent per year in the south through 2010 (Sam 1998), roughly in line with industrial GDP growth. This growth translates into Vietnam needing 400 megawatts of new capacity added per year to meet demand. Sixty percent of energy consumption currently comes from traditional fuels; however this is rapidly changing as coal, gas, and hydropower projects are developed. The energy options being considered by the Vietnamese government have potentially serious environmental impacts. Hydropower directly alters rivers and watersheds. Coal power has severe landscape and pollution impacts. Oil and gas have significant pollution impacts. And nuclear power looms on the horizon with a number of countries proposing projects in Vietnam.

The geography of industrial development also has environmental ramifications. As industry is overwhelmingly concentrated in and around Ho Chi Minh City and Hanoi, pollution and the urban problems associated with industrialization such as migration, crowding, and infrastructure deficiencies will be greatest in these cities. Uneven pollutant distributions may not only overwhelm the assimilative capacity of local environments, but may also unfairly affect those forced to live in the shadows of Vietnam's new and old factories. The rapidly developing industrial zones of Dong Nai, Binh Duong, and Hai Phong are fast becoming pollution "hotspots."

2.4 Government Efforts to Protect the Environment

In a direct response to these environmental problems, and to community complaints and media reports, the Vietnamese government has instituted significant changes in its legal and institutional framework for environmental management. As the Minister for Science, Technology, and Environment asserts, "Industrial pollution is starting to impact greatly on the social and economic development of Vietnam" (Interpress Service 2001). Since 1993, the government has promulgated an umbrella Law on Environmental Protection, standards on air and water quality, a decree on environmental fines and enforcement, a decree on the implementation of Environmental Impact Assessments (EIAs), and a long list of circulars and directives to advance environmental protection. These measures have created a system of national pollution standards and procedures for monitoring and enforcement. Table 2.3 presents an overview of environmental laws and directives related to pollution issues in Vietnam. The full text of the Law on Environmental Protection is provided in the appendix.

This combination of laws, decrees, and regulations forms the basis of a traditional "command-and-control" environmental regulatory system. The government has opted (and been supported by international donors

Table 2.3
Environmental Laws and Decrees

Name	Purpose	Date
Resolution No. 22/CP	"Resolution on the Functions, Responsibilities and Organization of the Ministry of Science, Technology, and Environment" establishing the National Environment Agency.	May 1993
Law on environmental protection	Umbrella law for environmental protection	December 1993
Decree No. 175/CP	"Government Decree on Providing Guidance for the Implementation of the Law on Environmental Protection."	October 1994
Circular No. 1420/QD-MTg	"Instruction for Guiding Environmental Impact Assessment to the Operating Units."	December 1994
Circular No. 1807/QD-MTg	"Regulations and Organization of Appraisal Council on Environmental Impact Assessment Report and Issuing Environmental License."	December 1994
TCVN 5937	Environmental Air Quality Standards	1995
TCVN 5942	Environmental Water Quality Standards	1995
Circular No. 715/QD-MTg	"Instruction for Guidance on Setting Up and Appraising the Report of Environmental Impact Assessment to the Direct Foreign Investment Project."	April 1995
Decree No. 26/CP	"Government Decree Providing Regulations on the Punishment of Administrative Violation of Environmental Protection Law."	April 1996
Circular No. 2433/TT-KCM	"Guiding Government Decree No 26/CP providing penalties for administrative violation concerning environment protection."	October 1996
Directive No. 36-CT/TW	Directive on Strengthening Environmental Protection in the Period of Industrialization and Modernization of the Country.	June 1998

and advisors) to develop a strategy that requires the state to implement comprehensive standards (both ambient and sectoral pollution standards), monitor compliance with these standards, and sanction noncompliance through fines and other punishments.

The state-centric, command-and-control nature of Vietnam's environmental regulatory system depends largely on the strength and ability of agency action. The state is expected to perform a number of key tasks, and to perform them well, in order to reduce industrial pollution. And as analysts have pointed out, Vietnam begins its rapid development "already carrying a heavy burden of excess population, depleted resources, and a seriously degraded environment. It has no margin for error, no unexploited cushion of resources on which to fall back" (Rambo 1994: 10).

In order to advance its command-and-control environmental system, the government has established a new set of institutions responsible for environmental and natural resource issues. The main agency responsible for industrial pollution issues is the National Environment Agency (NEA) within the newly created Ministry of Natural Resources and Environment (MONRE), which replaces the former Ministry of Science, Technology, and Environment (MOSTE).[2] The NEA has ten divisions including:

1. Administrative and planning;

2. EIA appraisal and environmental technology;

3. Environmental inspectorate;

4. Environmental education, training, and information;

5. Pollution control;

6. Nature conservation;

7. Networking and environmental database management;

8. Environment policy and legislation;

9. Environment monitoring and State-of-the-Environment; and

10. Environmental protection magazine.

The NEA has approximately eighty staff members working to implement environmental laws throughout the country.

At the local level, the government has established Departments of Science, Technology, and Environment (DOSTEs) in each province. DOSTEs are responsible for implementing environmental laws for all but the largest firms, and generally have monitoring and inspection teams. DOSTE staff working on environmental issues have doubled in recent years from 120 nationwide, to over 260 in 1998. Environmental departments have also been established inside other ministries such as the Ministry of Planning and Investment's Department of Science, Education, and Environment (DSEE), the Ministry of Industry's Technical Department, and the Ministry of Construction's Department of Science, Technology, and Environment (DOSTE). Nonetheless, Vietnam still has one of the smallest environmental staffs per capita in the region.

The Vietnamese government has also recently established a National Environmental Monitoring Network that collects data on ambient environmental conditions. This network consists of nineteen environmental monitoring stations spread across the country that monitor air, water, soil, noise, solid waste, and coastal environmental quality. These stations vary significantly in what they monitor, how they monitor, and staff capacities for analysis and reporting data. The government is working to expand this system. By 2005, MONRE hopes to have fifteen automatic air monitoring stations operating in Hanoi, Ho Chi Minh City, Haiphong, and Danang, and to establish mobile sampling capabilities at the provincial level. The Hydrometeorological Service of Vietnam also collects ambient environmental data through a network of twenty-two monitoring stations.

The primary forms of "monitoring" in Vietnam occur when the NEA and the provincial DOSTEs review Environmental Impact Assessments for industrial projects or conduct environmental inspections of industrial facilities. The DOSTEs and the NEA are responsible for reviewing and appraising EIAs written for industrial facilities and major infrastructure projects. EIAs unfortunately are not made public and there is no procedure in Vietnam for public review. The Ministry of Health similarly conducts assessments of the work environment inside factories. The Ministry of Agriculture and Rural Development has its own monitoring network for evaluating river water quality. And there are over 160 environmental organizations, associations, and research centers that collect information on different aspects of environmental quality in Vietnam.

The enforcement of laws in Vietnam often effectively occurs at the local level. The institution responsible for implementation of government policies in general is the People's Committee, and at the most local level, the ward- or hamlet-level People's Committees. In urban settings, neighborhoods are divided into wards (*phuong*) which have their own People's

Committee, with groups of streets (*khu pho*) organized on a more informal level. In rural settings, the hamlet (*thon*) is the lowest level of People's Committee. A group of hamlets makes up a commune (*xa*), which is represented by another People's Committee.

While still very new, it is surprising how little attention has been paid in Vietnam's laws and institutions to actual implementation and enforcement. Although significant efforts have been focused on establishing these programs, it appears that officials assume that with the laws in place, polluters will simply comply. Even the pollution permitting process in Vietnam until recently had no procedures for compliance schedules, or other processes or timelines for bringing polluting firms into compliance. Although noncompliance is the norm rather than the exception, enforcement remains almost an afterthought.

A recent, high-level directive (No. 36-CT/TW) indicates that the government may now be considering ways to better enforce existing laws (Communist Party [CP] 1998). This Politburo document recognizes that as "many cities and industrial parks are polluted by wastewater, air emissions, and solid wastes," it is necessary for Vietnam to take serious measures to promote environmental protection. The directive lists possible policy measures including: taxes and credits to encourage clean technologies; giving people access to information on the environment; supporting public participation in environmental protection; incorporating environmental considerations into development plans; promoting pollution prevention strategies; shutting down enterprises that seriously pollute; more effectively treating wastes; increasing research on environmental protection; and promoting broader international cooperation around the environment (CP 1998). Although this directive shows a high level of awareness about environmental issues, the critical question is whether these policies and plans are implemented, and how the government's pronouncements are translated into real incentives for action.

2.5 Challenges for Environmental Protection

While Vietnam has what some consider a strong central government, it faces problems (common to many countries) in its ability to implement environmental regulations at the local level. A lack of funds, trained personnel, and political influence constrain the effectiveness of environmental agencies. During a period of rapid industrialization and urbanization, local environmental agencies and planners are struggling to get a handle on even the most obvious sources of environmental degradation.

To date, the Vietnamese government has not made much progress on implementation of its new environmental laws. Institutional weaknesses in fledgling environmental agencies have led to very slow progress on monitoring and enforcement. Inspection divisions in the NEA and the DOSTEs are understaffed and poorly trained for their duties. And significant inconsistencies remain in the implementation of regulations.

Vietnam now has fairly strict standards on the books, but at the same time, generally weak enforcement of these standards. The government's inability to implement its laws has served to undermine the credibility of environmental standards and enforcement. For example, the current system of fines and penalties, which are extremely low and rarely enforced, motivate disdain more than fear from industry (UNIDO 1998). It is much cheaper to pay a fine than to change a production process or install treatment equipment. This failure to enforce standards occurs for both powerful foreign firms and state enterprises.

City governments similarly face difficulties in implementing environmental plans that have been developed over the last few years. Urban master plans often lack clarity and are difficult to turn into physical realities. One of the biggest challenges of environmental planning and policy in Vietnam has been the difficulty of trying to keep up with the rapid changes occurring in economic activities. Policy development and planning has not been flexible enough to respond to, or predict, emerging problems.

A deeper problem however, is that policies and plans related to environmental issues remain peripheral or subordinate to the dominant dynamics of industrialization and urbanization. Although environmental protection is now the focus of many laws and directives, the environment is still primarily treated as a resource or sink to be exploited. Planners in Vietnam seldom consider the environment's inherent value or longer-term importance to sustainability. This contradiction between environmental uses and protection is a major stumbling block to fully integrating sustainability concerns into development planning and implementation.

The continued failure (or more optimistically, the slow implementation) of traditional environmental regulations highlights the limits of simple command-and-control strategies for environmental protection in developing countries such as Vietnam. Standards are easy to promulgate, but they are costly and difficult to enforce. EIAs are easy to require of firms, but rarely actually change industrial practices. And punishments are easy to stipulate, but are very difficult to implement in the face of political opposition.

Promoting market-based mechanisms, which has become a mantra of consultants, conferences, and aid projects of late, appears to be problematic as well. Put simply, the institutions necessary to support market-based regulatory strategies do not yet function in Vietnam. The real-time monitoring required to support even a simple pollution tax does not exist. And again, the political will to impose large enough fines or taxes to make a difference for a polluting firm, is still lacking.

It would thus appear that Vietnam is destined at least in the near term to a low-road path of dirty industry, increasing pollution levels, unenforced environmental laws, and expanding social and ecological costs. The urban areas of Vietnam seem to be on their way to becoming the next Bangkok, Manila, or Jakarta.

Fortunately, there is evidence that existing processes at work in Vietnam may hold the potential to avoid this fate, and to effectively regulate industrial pollution. As I will explain, Vietnam is evolving a system of regulation that responds more to local mobilizations than top-down prescriptions. To understand these dynamics it is necessary to look more closely at specific cases in Vietnam, and at the central actors involved in these processes.

3
Cohesion, Connections, and Community Action

The people understand, the people discuss, the people implement, and the people monitor.
Vietnamese Communist Party slogan.

Contrary to stereotypes of communist countries, communities in Vietnam can be surprisingly strong and confrontational in their dealings with government agencies and firms. The community around Dona Bochang Textiles is a case in point. After researching the environmental effects of the factory, I set out to interview impacted community members. In another country I would have likely hopped on a motorcycle and driven out to the community. In Vietnam it is never so simple. The government works hard to control outside access to communities, requiring multiple permits, approvals, and often chaperones. After securing these approvals, I set off in a government Landcruiser (the vehicle of choice to attract the most attention), accompanied by a research assistant, two representatives from the Department of Science, Technology, and Environment (DOSTE), a police officer (to protect me?), two representatives of the local People's Committee, and a driver.

When we finally arrived at the first community member's home, my entourage included five government officials (including one in uniform). Within minutes of sitting down, twenty or so curious neighbors had gathered in the room, spilling out through the doors and windows. And now, with a gallery of close to thirty people, I was to interview a local family on the sensitive issues of the factory's pollution, community complaints, and the government's responses.

Despite the ridiculous conditions for the interview, the family was shockingly frank about local environmental problems and openly critical of both the factory management and government authorities. As the family expounded on their problems, it was impossible to miss the DOSTE chaperone scowling and taking careful notes of the criticisms leveled at his organization. But the family showed little fear. Mr. Nguyen (not his real name) carefully detailed the environmental and health impacts of the factory's pollution on the neighborhood. He complained that the government and the Taiwanese factory managers had ignored their complaints for years. A neighbor alluded to corruption blocking solutions. Other community members joined in on a free-wheeling critique of the firm and the government.

To an outside observer it is surprising that community members in Vietnam would openly criticize local government officials, particularly in front of a foreigner. And it is even more surprising when these complaints turn into community actions against a factory (particularly one like Dona Bochang that is partially owned by the government). Many analysts assume that civil society in Vietnam has been crushed under the weight of the Communist Party. "Stalinist" central-planning and decades of war have made the Vietnamese state famous for governing with a heavy hand and seemingly ignoring local concerns about economic, social, and environmental issues. This stereotype of all-powerful communist rule, however, oversimplifies the complexities of Vietnamese politics.

One of the surprises of this research has been that communities and civil society have not been entirely repressed by the Vietnamese state. Despite state control, substantial community action and bottom-up pressures for reform exist throughout Vietnam. Communities can be vibrant, aggressive, and sometimes extremely well organized. High levels of community participation and bottom-up political pressures have played important, though varied, roles in state policies. Although these processes are complex and often conflicted, they hold significant potential for social, political, and environmental advances.

The communities I studied, of course, differed in their responses to environmental problems, in their abilities to mobilize, and in their effectiveness in demanding environmental improvements. A number of key factors influenced the nature and success of community actions including: internal cohesion, external linkages, and mobilizing strategies. This chapter examines these dynamics, analyzing specific community mobilizations, how and why they occur, and the conditions under which they are effective. The case studies provide examples of communities that both do and do not mobilize effectively, offering a window on the processes of mobilization in Vietnam, and the characteristics of effective community pressures for environmental protection.

3.1 Community in Vietnam

Communities, while central to this study, are often difficult to define both physically and sociologically (Wilkinson 1986). In countries such as Vietnam, this is compounded by challenges of accessing reliable data on community demographics. The community studies literature defines community quite broadly and in often varied terms (Bell and Newby 1972). Three common clusters emerge however: definitions based on fixed and bounded territories; definitions based on shared social relations within a territory; and definitions based on shared identities (Hillery 1955, Johnston, Gregory, and Smith 1994). I build on these frameworks to define community as a social network of interacting individuals concentrated in a bounded territory.

The communities I studied were defined by their physical, environmental, and social ties to specific locales, and often to specific factories. The case study communities discussed here are thus the physical groupings of people connected to the case study factories. The communities are spatially connected, but also have interests, identities, and interactions in common.

Few people in Vietnam have the resources or freedom to move. This connection to place creates a fundamental interest in the quality of life and the environmental conditions of the community. Community members realize that they have shared interests in the provision of collective goods such as clean air and water. And they often come to realize that the provision of these goods depends on the efficacy of their collective actions. Community is also a source of less tangible supports such as social assistance and political support to respond to adversities. In Vietnam, village and kinship ties are particularly important. Social networks linking individuals and households in a specific locale are difficult to define, but are critical to the functioning of a community.

The traditional community in Vietnam is based around the hamlet or village. Hamlets are closely clustered groupings of households that may include people "considerably differentiated in wealth and status yet linked together by extensive social ties within a communal framework" (Luong 1992: 55). Hamlets have historically represented territorial and voluntary associations based on genealogical and kinship ties. Luong explains:

> Villages maintained sharp boundaries through villagers' participation in the collective worship of tutelary deities, an extraordinary degree of village endogamy, the institution of communal land, and their nucleated settlement pattern. . . . Correspondingly, in each northern and central rural community, "insiders" and "outsiders" among village residents (referred to as *noi tich* and *ngoai tich*) were sharply distinguished. It might take three to four generations for the descendents of outsiders (i.e., those not born in the community) to gain the full membership status of *noi tich*. . . . The categories of *noi tich* and *ngoai tich* constituted part of the general and sharp distinction between the members and nonmembers of a social unity, be it kinship, communal, ethnic, or racial. (Luong 1992: 229)

Villages are much more than just clusters of homes. As Jamieson (1993) explains, historically villages were closed, corporate communities with a high degree of autonomy, physically and symbolically set apart from other villages. Villages owned land, which was redistributed communally every three years, and were responsible for paying taxes, contributing corvée labor, etc. Villages have thus assumed many roles and responsibilities that individuals might be responsible for in other cultures. The village provided for, and protected the individual households within it. And as Jamieson notes, "So long as it met its obligations, the village was left alone" (1993: 30).

Households—the building blocks of Vietnamese communities—have their own political economies (Douglass 2001). Politics within the household, usually based around age and gender, influence internal decisions and external connections to other households and the community at large. Even with the rise of collectivism and increased female participation in public activities, men have continued to dominate both the communal system and kinship relations. These gender dynamics also influence responses to environmental issues (Agarwal 2000). For instance, women are often closer to environmental concerns than are men, as they are responsible for providing water, fuel, and so forth for their families. However, if women do not have the time, resources, or power to participate in collective action regarding the environment, households may not participate at all.

Broader family relationships and commitments have historically served as models for social organization in Vietnam. As Jamieson (1991: 10) explains, the "Vietnamese believed in a universal order with pervasive moral justice. However, the scales of justice were balanced not for individuals but for groups, especially the family." Interactions within villages reflected these values as well. Competition for status in the village was based around redistribution of wealth. "The basic principle of redistribution could be summarized as 'from each according to his desire for face, to each according to his willingness to lose face'" (Jamieson 1991: 13).

Communities in Vietnam also have a long history of resistance and rebellion. To grossly simplify, the Vietnamese resisted the Chinese for 1,000 years, the French for 100 years, and the United States for just over ten years. As Marr argues, "The continuity in Vietnamese anti-colonialism is a highly-charged, historically self-conscious resistance to oppressive, degrading foreign rule. Possessors of a proud cultural and political heritage, many Vietnamese simply refused to be cowed" (Marr 1971: 4, cited in SarDesai 1992). Resistance to oppression continued after the French were defeated in North Vietnam. As Kolko notes, Communist Party leaders "repeatedly had to concede to mass pressures in the hope of guiding and controlling them, for they needed the people, whose force and enormous sacrifices alone carried the Party to victory in 1945, 1954, and thereafter" (Kolko 1995: 5).

Communities, of course, are porous institutions. Members join and leave. Issues bring people together and drive them apart. Boundaries expand and contract. Communities are thus shifting and contested terrains (Fortmann and Roe 1993). They are both simple territorial bounds, and complex social organizations that are continually redefining themselves. The concept of community can also be an aspirational goal. People desire a "sense of community" in which they experience membership, influence, integration, and shared emotional connections (McMillan and Chavis 1986).

A single community can also be fragmented along multiple lines. Communities often include layers of power and conflicting interests. Individuals within the community play different roles and advance different interests. The most obvious differences pertain to income generation. The case study communities discussed here include people who make their living as government employees, industrial workers, farmers, petty traders, and factory managers. These groups vary politically, economically, and socially. Some are members of the Communist Party, others are politically disenfranchised; some are members of religious organizations, others belong to mass organizations; some are rich, others are extremely poor.

Even a single individual may have multiple roles and interests relating to the community, or more specifically to industrial activities in the community. A person may act at one time as an income generator, and at other times as a parent, citizen, or neighbor. Individuals clearly have different uses for the environment as well. For instance, near the Tan Mai factory, which I discuss next, a man who earns his income collecting paper fibers from the factory's wastewater simultaneously complains to the government about the factory's wastewater. Workers near Viet Tri Chemicals tell similar stories of conflicting concerns and interests.

Despite these complexities, communities are an important unit of analysis for environment-development debates. As Evans (2001: 14) points out,

looking at communities focuses attention on the politics of collective action among households with connections to one another. Communities build identities based on geography, history, and shared adversity. Their members share life chances. They are vulnerable to the degradation of the places to which they are attached. Talking about "communities" enables us to connect livelihood struggles of ordinary citizens to issues of sustainability while retaining the critical insight that these are not simply individual battles but always have an element of collective contestation.

The people at the heart of the case studies are not just rational actors making individually self-interested decisions. And at the same time, they are not simply an undefined "collective" driven by Vietnamese culture or tradition. Community members are engaged in social processes through their families, neighbors, and local political institutions. And these community members are most successful when they come together to mobilize strategically.

3.2 Cases of Community Mobilization

Vietnam continues to have surprisingly active communities when it comes to environmental concerns. As the World Bank (1997: 83) has noted, "the level of awareness of environmental pollution is high in Vietnam as compared to other countries at comparable stages of development." This may be due in part to Vietnam's higher literacy rate, and to the fact that Vietnam has been able to observe the problems caused by rapid development in neighboring countries. Nonetheless, the cases show wide variation in levels of community awareness and mobilizations around environmental and health concerns. Communities in Vietnam respond differently to pollution problems and differ in their effectiveness. Three key differences become clear through analysis of the detailed cases: communities vary in their internal cohesion, external linkages, and mobilizing strategies.

3.2.1 Pollution in the Pews: Dona Bochang Textiles

Situating the Dona Bochang Textile factory in the middle of a residential community is a glaring example of unplanned urban development and the absence of zoning in Vietnam. The factory and the local residents are separated by no more than a three-meter-high wall and a dirt road that runs along the perimeter of the factory. People live cramped together in small houses along the back wall of the factory; the local Catholic church is along another wall. The factory's air emissions, when not blowing into the residential area, blow directly into the church. Unfortunately for the parishioners, the church was, until recently, an open-air building with little more than a roof, an altar, and rows of pews.

Local regulators say the air pollution from Dona Bochang is really not all that bad. The neighboring community, however, does not seem placated by the thought that other communities have it worse. In their view, pollution from the Taiwanese-run factory is a continuing assault on the neighborhood, affecting peoples' daily lives, disrupting special occasions, even defiling their center of worship. Pollution impacts such as respiratory problems, corroded roofs, and blackened plants have led to an escalation of community actions that have included writing regular complaint letters, working with the media, throwing bricks at the factory, and developing a long-term campaign to either make the factory a better neighbor or move it altogether.

The community around Dona Bochang is on the middle rung of Vietnamese urban communities, certainly not rich by western standards, but much better off than nearby rural communities. The community has prospered under the *Doi Moi* reforms. Many families run their own businesses—either producing simple products such as wood furniture or trading petty goods. A number of people in the community work in nearby industrial zones, adding an important source of income to their households. What is more surprising perhaps is how many people in the community choose not to work in these factories, as they can earn a higher income than the minimum wage of $40 per month without doing factory work.

As the Taiwanese managers soon found out, the people living around Dona Bochang are a tightly knit community. Approximately 90 percent of the community are Catholics who moved to the area in 1954, fleeing the Communist victory in the north. As Catholics, they were first protected and received special status from Ngo Dinh Diem, the former president of South Vietnam (and himself a Catholic). After unification under the Communists in 1975, Catholics were politically marginalized. However, throughout Vietnam's political changes, they have retained a distinctive identity in an overwhelmingly nonreligious country.[1] The community's solidarity and internal social capital has been strengthened by more than forty years of church organizing.

The church's location next to the factory has made it a critical base for discussions on the impacts of the factory, and for sharing grievances. Although the parish priest went out of his way to make clear in interviews that he was not organizing community members around environmental issues, he did admit that the church had become a center of discussion for these concerns. The priest (personal interview—6/7/97) explained somewhat reluctantly,

One time we had an official meeting regarding the pollution and the problem of community members throwing rocks at the factory. The Catholic people come here to talk to me about the pollution all the time.

We have a diplomatic relationship with the factory. I have met with the General Director of the factory. I recommend that they send letters to the *Phuong* [ward] People's Committee. And I also tell them to be patient. It takes time to solve problems.

While supportive, the church alone is not the leader of the community's actions. In fact a number of people intimated that the priest was afraid to get involved in these issues and thus often told people to work through the People's Committee. At least one community member asserted (Ms. Lan—6/5/97) "the priest wants to avoid conflict." This makes some sense, as Catholic priests have been the subject of repression themselves.

The real base of community mobilization around Dona Bochang is an informal social network that exists among community members. People live, socialize, worship, and sometimes work together (a number of people in the *khu pho*—the lane along the back of the factory—make similar furniture for a living). This community is also cohesive partly because of its past experiences of religious discrimination. So although Catholics may be more fearful of state oppression (and thus less likely to complain), they are also more unified and cohesive in their demands when they do make them.

There is clearly a very high level of community interaction in the neighborhood along the back wall of the factory. Each time I went to interview a family, other neighbors would quickly join in the conversation. People walked freely into each others' houses. The lane abutting the factory is the de facto playground for the neighborhood children. There is a real "sense of community" in the neighborhood.

In my interviews, one family stood out as leaders of community action. They seemed fairly well educated and quite well off for the community, running their own small household enterprise finishing wood furniture. Living and working just a few feet from the factory wall, the family had collected a thick file on the factory's pollution, including press clippings, letters they had sent to various government agencies, the responses they had received, and photographs of pollution impacts. They regularly drafted letters for others to sign. They had been on official delegations to the factory and to government meetings. They had even made a video of the pollution. In many ways, they seemed fearless in their quest to end the pollution, a quest they have yet to finish.

Community members have taken a wide range of actions to pressure Dona Bochang. One community member explained (Ms. Lan—6/5/97),

We have complained many, many times to the DOSTE, the newspapers, the TV. Many times. We have also complained to the factory many times. We had a meeting between the people, the factory managers, and the People's Committee in the *Phuong*. We have also had many meetings of our own. People around the factory have done many things. Young people have thrown rocks at the factory many times. Because of the rock throwing, the factory went to the *Phuong* People's Committee to try to solve the problem. But still the factory didn't change and the people still opposed the factory. The factory has even tried to hire the young people who threw the rocks and complain. Because everyone knows each other so well, we can have coordinated actions together. We have so many things in common.

After several years of having their complaints ignored, the community was ignited by an incident in 1993. On the day of a local wedding, pollution from the factory coated trays of food laid out for the reception in a layer of black soot. A group of community members considered this the last straw, marched to the front gate, and threatened to tear down the wall and shut down the factory if the manager did not come out to negotiate with them. A number of young people threw bricks at the factory to make clear their anger.

A factory representative did finally emerge and asserted that the company was doing all it could and promised the problems would be solved. The community forced the manager to sign a statement attesting to the level of pollution. Photographs were taken of the pollution impacts. Several months later, when nothing had changed, the community brought their complaints, the pictures, and the signed statement to the Dong Nai Department of Science, Technology, and Environment (DOSTE) and then to the media. Figure 3.1 gives a sense of the pollution blowing into the community.

After newspaper reports questioned the failure of the government to regulate the pollution, the DOSTE agreed to take action. It responded to the community complaints by organizing an inspection team and several meetings between community members and the factory. The community however, criticized this inspection process, charging that because it was a planned inspection, the factory would be able to turn off the polluting equipment before the inspectors arrived. Community members argued that their daily experiences were more accurate than the data collected from the inspection. Later, when pollution levels resumed, the community sent more written complaints to the government and the media. This renewed pressure motivated more meetings and finally resulted in the factory agreeing to install equipment to reduce its emissions.

By the fall of 1997, the neighbors of Dona Bochang had achieved a qualified victory over the factory. Since the wedding party incident, the factory had made three changes to reduce its air pollution. First, it built a taller smokestack—the classic solution to local environmental problems. When that did not reduce the local impacts, the factory changed its practice of "blowing the tubes" from its boiler (where built-up soot is forced out of the smoke stacks), which was a major source of the black soot spewed on

Figure 3.1
Air pollution from Dona Bochang

people. Finally, when the problems were still not resolved, the factory installed an air filtration system to capture the pollution. This process took several years, but resulted in a significant reduction in air emissions (according to the firm and the Dong Nai DOSTE). The community also agrees there have been improvements, but continues to pressure for further pollution reductions.

Even with this significant and sustained pressure the community has felt frustrated by their progress. A community leader asserted (Ms. Hao— 8/6/98),

Since our last letter to the DOSTE, we have met with staff at the DOSTE. They sent a letter to the National Environment Agency (NEA), and then the NEA instructed the factory to prevent pollution in three parts of the factory. By the end of last year, the factory was supposed to reduce or treat wastewater and air pollution. However, it still doesn't seem like they have done much.

The factory receives help from higher people in Dong Nai so they don't really care about citizen complaints. The People's Committee of Dong Nai protects the firm. We think the company has a close relationship with some VIP in Dong Nai. But nowadays people are not afraid to tell the truth about problems. Even if a policeman does something wrong, the people will complain.

We don't need compensation for health effects. In fact, we don't want to get money and still suffer from health problems. We want to move the factory to an industrial zone or improve the factory. We want to get the company's board of directors to come out here to breathe as we do.

We believe if there is collaboration from all sides we can reach the objective. If government authorities are serious about the environmental law, they can solve the problems.

Although the community is not satisfied with the changes made thus far, Dona Bochang has been forced to take further steps to reduce its air pollution. The company recently added caustic soda to the scrubbing liquor in their cyclone to capture sulfur dioxide (SO_2), increased the water flow in the cyclone to reduce soot, added an ammonia scrubber, and finally changed their procedures for "blowing their stacks," which was the main cause of the large black clouds of soot spewing into the community. The factory now cleans their boiler once per week instead of once per month so the soot does not build up as much in the stack. The company claims to have spent 200 million dong (roughly $17,000) on the air treatment equipment for the boiler alone. The company, however, has not installed a wastewater treatment plant (expected to cost over $3 million) that the community has been demanding and that the government has officially required.

The community has thus only been partially successful in winning increased environmental protections. Complaints have succeeded in getting the factory to invest modest amounts of money in cleaning up the air pollution, but they have not been able to motivate the factory to invest the millions of dollars needed for comprehensive environmental protection. The community has also never received any compensation for the pollution. Even the people whose well water is too polluted to drink have never received compensation. Figure 3.2 shows the leaking wastewater canal that flows from the factory through the community.

The state's role in this case is complicated. As the factory is a joint venture, the Dong Nai People's Committee owns 12.5 percent of the company. Community action is thus in conflict with the short-term economic interests of the provincial People's Committee, which controls the DOSTE. The community members' perception that they had to overcome this conflict of interest led them to extralocal actors such as the National Environment Agency and the media to help address their problems. It also strengthened the community's resolve to keep pressure on the factory and on the provincial authorities. Community members did not trust the state to take action without repeated pressure. At the same time, the fact that the factory is majority foreign-owned may have worked to the community's advantage: Vietnamese government agencies appear sensitive about public perceptions that the state is privileging foreign capitalists over common people.

DOSTE inspections of Dona Bochang have been conducted through an "intuitive" method, meaning no sampling was actually conducted. The factory is informed beforehand about when inspections will take place, and has time to alter its operating practices to reduce visible pollution. When the inspectors do arrive, they report that "air pollution from the boiler is negligible," "we did not see smoke on the day we visited," and the "white smoke from the boiler is just steam." Community members play a critical role in these situations in "monitoring" both normal operating conditions when the inspectors are not around, and evaluating the veracity of state inspections. Community monitoring techniques appear to be as good, if not better than the DOSTE "intuitive" method. The community takes pictures and videos of what it sees, and records health and other impacts from pollution incidents. Even a factory manager at Dona Bochang admitted this weakness in state monitoring, saying (Mr. Luat—

Figure 3.2
The wastewater canal for the Dona Bochang textile factory

4/11/97), "The DOSTE does not know in detail how much pollution there is. [Our response] mainly comes from the complaints of the people that motivates us to change. Each time the director general comes to the plant from Taiwan, the first question he asks is what complaints have we received. He takes environmental issues very seriously."

Despite conflicts, the community has had quite good relations with local government officials at the *Phuong* (ward level) People's Committee. The chairman of the People's Committee is a Catholic and lives near enough to the factory to know first-hand about its pollution problems. The chairman seemed accessible and concerned about the pollution problems, but also quite circumspect about his power to affect change in a foreign joint-venture factory. Because of his social ties to the impacted community members, he appeared more vulnerable to social pressures, and thus more responsive.

The chairman of the local People's Committee asserts that the community has more influence over the factory than he does (Mr. Viet—6/7/97), "We cannot fine Dona Bochang. We can only contact them and make suggestions about issues to be resolved. The company is really afraid of bad publicity. The company does things to appease the peoples' complaints." Another government official agreed, explaining (Mr. Het—10/11/97), "Peoples' complaints have a great impact and they play an important role. The complaints force factories to take action, and they raise environmental issues for companies built before the Law on Environmental Protection. Whenever there are protests, the company [Dona Bochang] does something to appease the people. But then they stop. When there are more protests, the company makes one more change. They don't really want to solve the environmental problems."

So in the end, the community around Dona Bochang has been able to mobilize its members, build linkages to the state and media, and conduct its own "monitoring" of pollution and inspector responses. This has led to limited successes against a recalcitrant and well-connected foreign firm. Many of the improvements at Dona Bochang are thus due to the strength, organization, and persistence of the community. However, it should be noted that the community on its own was not able to change the company. Letters and meetings with the factory owners did not result in pollution reduction. Success came through the community exerting pressure on local and national government agencies both directly and through the media. The community's linkages to local government officials and outside reporters were as critical as its own internal cohesiveness to its ultimate success in motivating state and corporate action.

3.2.2 Living Off Pollution: The Divided Community around Tan Mai Paper

Just meters beyond the outer wall of Tan Mai paper mill, a thriving industry exists in the shade of coconut trees. In ponds where rice fields used to lie, local villagers stand chest-deep in wastewater from the paper factory. Young men strain to lift nets out of the ponds, filled to the brim with the catch of the day: paper fiber emitted in the mill's wastewater.

As one part of this community literally lives off wastewater, selling recovered fiber to low-grade paper makers in nearby Ho Chi Minh City, other people pay the price in damaged crops, polluted drinking water, and dead fish. The Tan Mai case is an example of a divided community that both depends on the factory's pollution for income and is injured by its activities. Some community members work in the factory. Others complain of losing an entire year's crops with no compensation.

Although Tan Mai had been causing pollution since the 1960s, it was not until the factory increased its production in 1992 that community members organized as a group to demand recourse for dead fish and damaged crops. Between 1992 and 1996, community members wrote letters to the DOSTE, the media, and the factory management. The DOSTE investigated the claims of the community but never showed the results to community members and never awarded compensation for lost crops or fish.

Most people in the province accept that Tan Mai causes serious environmental impacts. The factory managers themselves acknowledge that they need a new waste treatment system. Even the people who make their living recovering paper fiber from the wastewater express their concern about the impacts of the factory's pollution. Local farmers cannot eat the rice they produce, using it only as feed for their pigs. Community members complain of nausea from air pollution; undrinkable well-water; nose, eye, and skin problems; and lower yields from their fruit trees.

One community member's story is representative (Mr. Nam—6/3/97), "Before I grew rice here. It was damaged by pollution so I had to switch to raising fish, but the fish also died. In 1995, I lost 350,000 dong worth of fish. After the fish died I decided to grow vegetables, but even the pigs couldn't eat the vegetables because of chemicals, so now I can't grow anything. How can we return to growing food?" Another community member explained, (Mr. Thanh—6/13/97),

In 1992, Tan Mai expanded the capacity of the factory. At that time the wastewater increased so the wastewater was too much for the pipe, and it flooded the land. When this happened, the roots of the rice became black. Before we could grow three crops per year. Now we can only grow one time per year. The yield now is five tons per hectare per crop. In the past we had 15 tons per hectare per year. . . . The DOSTE came to inspect and determined there was no pollution in this area. They took samples, but we never saw the results. They just told us that it was within standards. . . . I am sure the crops were damaged by pollution. I don't believe the results of the DOSTE testing. . . . We are worried about toxics in the rice. I don't eat this rice. It is only for feeding to animals. The roots are black, and the rice is weak. It cannot even stand up straight.

Even the people who collect paper fiber out of the factory's wastewater recognize the hazards of the pollution. One man explained how he had been collecting the fibers since 1985 (Mr. Qui—6/3/97),

I was the first person to start collecting *bot giay* (paper fibers). At first the factory complained about my work, but later they decided it was good for them and the environment. Otherwise everything would go to the river. They don't support me but they don't oppose me. They don't say anything.

I am not concerned about the chemicals in the wastewater although I stand in it up to my waist each day. But I am concerned about drinking water. I can't drink the water from my well. In this area, only 2 out of 10 wells have drinkable water. . . . I have complained about the pollution even though I make money from the pollution. I complain because the pollution affects my children. I am worried about the water quality and the smells. I sent a letter to the DOSTE and I have signed other letters.

I believe the costs of pollution are higher than the benefits. I make about $200 per month from the *bot giay*. But the factory loses a lot of money. They can't make any profit. Because this factory belongs to Hanoi, the factory has a lot of power and they are not afraid of provincial authorities.

The community around Tan Mai is both physically and emotionally divided. One group of families lives next to the factory's back wall, collecting the paper fibers; another group grows rice in fields nearby; a third lives in company-built apartments on the urban side of the factory; and a fourth lives in fish-raising houseboats on the river into which Tan Mai discharges its wastewater. Figure 3.3 shows one of the houses that has flooded its land

Figure 3.3
Waste fiber recovery outside the Tan Mai paper mill

to collect wastewater and paper fiber. The *Phuong* People's Committee has a young and dynamic chairman, who is quite open about the environmental impacts of the factory on the community and equally open about his frustration with not being able to change the situation. Through this local official, the community has submitted formal complaints to the factory and to provincial authorities. But as he explains, "The people in this area have children working in the factory. They can use electricity and water from the factory. So of course there are losses and benefits from the factory, so they don't want to complain too much" (personal interview—June 6, 1997).

Perhaps because of its divisions, the community lacks leadership to take action on the pollution outside the official People's Committee structure. Attempts to get community members to even sign letters have been frustrated. As one community member explained (Mr. Nam—6/3/97),

Pollution impacts all of the families in this area. They all have many complaints but no one wants to write a letter. The other villagers are afraid that if they complain the authorities are unhappy about them. People don't want to be noticed by the authorities. . . . There are several opinions in the community. Some families have someone working in the factory, so they are afraid to complain about the pollution. The other is the people who get benefits from the waste, and they don't want to complain either.

Tan Mai is owned and managed by central state authorities and is at the same time under the regulation of the National Environment Agency. Either through corruption or a concerted policy, the state has worked to block criticisms and demands for environmental improvements at Tan Mai. For instance, after complaints from the community, the DOSTE took measurements of water pollution at Tan Mai, but they were taken in a way that covered up the real pollution levels: some samples were actually taken upstream from the factory, where the water was relatively clean. The DOSTE then issued a formal memo stating that the factory was in compliance with environmental standards. Virtually everyone involved in this case (including government officials interviewed) recognize that Tan Mai is nowhere near compliance with environmental standards, yet this document is now accepted as "proof" of Tan Mai's performance.

The community also lacks ties to government authorities. Tan Mai Paper is a centrally run state enterprise. So while the *Phuong* People's Committee chairman seemed responsive to community complaints, the community was unable to advance its interests at the central government level. A local government staff member explained (Mr. Nghia—6/6/97) "The *Phuong* has no power. We cannot fine Tan Mai. We cannot close the factory. The *Phuong* only collects complaints and information from the people and sends it to the appropriate authorities. Because Tan Mai is controlled from Hanoi, the province cannot do much."

Community members simply seem beaten by the factory and state inaction. As one farmer explained (Mr. Huu—6/4/97), "I have never complained. I am a grassroots person so any complaints will have no effect. I have a low position and I cannot affect the authorities." Another shook his head and worried, "I don't think the environment will get better until the problem of corruption is solved." Community members have thus resigned themselves to the factory's continued pollution, admitting that they no longer write complaint letters as "they have no effect," "they only result in DOSTE coming out, measuring, and then disappearing," and "they get you noticed by the authorities." This discouragement is not uncommon. Other communities I studied also feared that complaints would be ignored or cause more trouble than they were worth. Nonetheless, some persevered and were sometimes successful.

The community around Tan Mai however, has been unable to overcome internal divisions and fears to mobilize against the factory. The community is endowed with reasonable capacities, including a mix of educated young members and industrial workers. It even has some connections to local government representatives. Nonetheless, it has not been able to forge broader state or media linkages, and its internal divisions have weakened its ability to pressure environmental agencies to take action against the centrally managed factory.

Then again, perhaps more cohesion would not have much effect. Tan Mai is, for a number of reasons, an extremely well insulated company. The government has targeted the paper industry for expansion and is aggressively promoting the three largest pulp and paper mills in the country, including Tan Mai. Promotion and protection of Tan Mai thus wins out over other interests (even tax collection), and blocks local regulation of its pollution. The firm in this case has such strong linkages with the state that virtually no amount of local pressure may be able to motivate stricter regulation. Recognizing this, community members have given up even submitting formal complaint letters.

Tan Mai is a classic example of a divided community, facing an unresponsive state agency, that has given up out of frustration. As one community member explained (Mr. Truoc—6/4/97), "The people are tired of complaining. The DOSTE or the People's Committee comes here, the factory reduces pollution for several months, then they pollute again." It may be that this Hanoi-run factory would not have listened to even the most mobilized community. But the internal divisions within the community guaranteed that they could not overcome factory or government resistance to change.

3.2.3 Lam Thao's Bitter Tea: The State as Polluter

A woman in her 60s led me and a group of neighbors to the wastewater canal they said was the source of many of their illnesses. With the skill and strength of a lifelong farmer, she dug down several feet into the soil to expose a leaking pipe. Farther on, several men used a crowbar to pry open the cement cover of the wastewater pipe to show more leaks and to illustrate how the wastewater had literally burned away the cement cover. The wastewater is so acidic it often has a pH of 1, and has leached into the community's drinking water supply. Even their tea, the lifeblood of the Vietnamese day, is now bitter. The positive side, one woman joked, is that they don't have to add any spices to make sweet and sour soup.

Unfortunately, acid in the wastewater is only one of the environmental problems faced by the communities living around the Lam Thao fertilizer factory. Lam Thao, built with Russian assistance in the 1960s, stands as a looming example of disregard for the environmental impacts of industrial development. Air pollution from the factory's production of sulfuric acid and superphosphate fertilizer carries sulfur dioxide, sulfuric acid, hydrochloric acid, hydrogen fluoride, and other toxins to the surrounding villages.

Although community members complain that they have been living with the pollution since the factory's opening in 1962, they assert that the pollution increased substantially in 1992, after the plant was expanded. In one hamlet, the People's Committee chair explained "We did not realize [how bad the pollution was] until the disease rates became high and some environmental organizations investigated." He asserts that the death rate in his hamlet doubled in 1992, the year the factory increased its output (personal interview—4/18/97).

Other health problems associated with the factory's pollution include "swollen skin, rashes, people losing teeth [they become loose and fall out from drinking the water]. The rate of cancer has increased recently, and children have problems with their throats." A local nurse notes that air pollution has caused high rates of lung illnesses, and "swollen throats are very common close to the factory" (personal interviews—4/18/97). Pollution from the factory also regularly damages crops. Water and air pollution kill rice, banana trees, and coconut trees. Fruit trees that survive gradually decrease their yields. Accidental leaks from the wastewater pipe regularly destroy nearby farmers' crops. One commune alone claims to lose 300 million dong (approximately $22,000) worth of rice per year to pollution damage. Figure 3.4 shows damaged banana trees from air pollution from Lam Thao.

Communities around the factory have no problem assigning blame for the health and economic impacts they face. However, they differ in their strategies for bringing Lam Thao to task for its actions. There are two distinct communities near Lam Thao that are affected by the factory's pollution. Community A is directly adjacent to Lam Thao and is primarily affected by the factory's water pollution. Community B is located across the Red River from the factory and is primarily affected by the factory's air pollution. Despite similarities in terms of community cohesiveness, differences in capacity and linkages strongly affect each community's ability to challenge the factory's environmental degradation.

The members of the community next to Lam Thao—Community A—are better educated than their counterparts across the river; they hold regular meetings and have applied direct pressure on the factory, including letters, demands for meetings with factory managers, and complaints to government authorities. Some of the workers from the factory live in Community A, so the community has divisions in opinions and alliances on the pollution issues. Nonetheless, Community A has been able to mobilize community pressures and to tap into existing ties to local government officials. Several of the leaders of the community were formerly employed in the district and province People's Committee. Houses around the commune also proudly display certificates declaring their owner's status as war heroes, which helps in negotiations with government and Party officials.

Community B, on the other hand, is a much poorer commune, made up of subsistence farmers, with no electricity in their hamlet, a high incidence

Figure 3.4
Damaged banana trees from air pollution from Lam Thao Superphosphates

of child malnutrition, and few connections to the hierarchy of the district's People's Committee. Community B has very weak leadership, seems docile in the face of even the worst environmental insults, and has only been able to win small compensations each year, which community members admit pay for only about 5 percent of their crop losses.

Community B is actually more tight-knit than Community A, with less migration in or out of the community, and a higher percentage of people working as farmers. However, despite some divisions in Community A, it has been much more successful in pressuring Lam Thao to change. Community A is also much more sophisticated about its legal rights, recognizing the power of the Law on Environmental Protection in guaranteeing a community's right to be protected from pollution or at least compensated for its impacts. Community A benefits not only from its sophistication but also from its greater linkages, in this case its more developed ties to provincial authorities. Community B, on the other hand, is both physically and politically cut off from provincial decision makers. Both communities rely on the leadership of the local People's Committee chairman to press their cases. The chairman of Community A, however, is much more capable at pressuring higher state authorities and factory managers.

Based largely on the complaints of Community A, Lam Thao has made some important changes in its production practices. They first hired technical researchers to conduct an Environmental Impact Assessment and a waste audit of the factory in 1995. In the following years they made a number of the changes recommended, including: installation of a wastewater treatment system, a dust treatment system, and a circulating pond and neutralization system (using lime) for wastewater; covering the wastewater canal with a cement cover; transitioning to a wet method in superphosphate production; and replacing pyrite with sulfur. The total cost for these changes was 8.5 billion dong ($700,000) in 1996 and another 30 billion dong ($2.5 million) in 1997. The factory now also conducts air monitoring (for sulfur dioxide, sulfur trioxide, and fluorine) two or three times per shift. They also continuously monitor pH and check biochemical oxygen demand (BOD) of their wastewater three times per year. Community A has also been successful in winning compensation from the factory for crop damage caused by the pollution.

The factory managers assert that they first became aware of environmental issues in 1992 based on health impacts on workers. They were then alerted to the impacts on community members by local complaints. One manager noted (Mr. Tuyet—10/13/97), "In the past we didn't have any community pressure because awareness was low. Now people know how to complain. If we have any problems, they complain immediately." A more senior manager asserted that the company is taking the complaints of the community to heart (Mr. Loan—5/27/97), "We know it costs a lot of money, but otherwise we will be closed. The community will complain and will force the government to close the factory, or will organize a protest in front of the factory."

Community A has taken a range of action to motivate these changes. A community leader explained (Mr. Duoc—5/27/97),

We first complained in 1992. We recognized the environmental problem a long time ago, but there was no organization working on environment, so we just complained to the district People's Committee. They did not take it seriously. After the Center for Environmental Management was established we could complain to them. . . . We wrote letters to the district People's Committee and to Vietnam Television. Vietnam Television visited. Other regions are benefiting from the products of Lam Thao, and we are suffering.

Another member of Community A explained (Mr. Do—4/17/97),

The response began at the grassroots level. People complained to the commune People's Committee. The commune People's Committee wrote letters to the district and province People's Committees. Some individuals wrote letters to various organizations. . . . The most important factor is the commune pressure on the factory. Whenever we complain we base it on the environmental law. . . . We have clear evidence. . . . When crops are damaged, people complain to the People's Committee. The People's Committee reports to Lam Thao. Lam Thao then sends people out to review the complaint. Then we meet to agree on the compensation. The highest compensation for our commune has been 25 million dong. . . . We have also talked to the factory directly many times. We have had many community meetings to discuss pollution. People only really complain, though for major pollution cases. . . .Lam Thao has a monopoly so they don't care much about quality or price. Community pressure is thus the biggest pressure. . . . There needs to be a mutual understanding between the community and the factory. The community understands Lam Thao needs to make a profit. And Lam Thao understands they sometimes hurt the people nearby. From this understanding they can agree on compensation and actions.

A leader of one of the affected hamlets in Community A explained (Mr. Binh—4/17/97),

We have serious disagreements with the factory. We set up appointments with the managers but they often won't meet with us. Sometimes we ask the People's Committee to go to Lam Thao to meet with them. In March, after failing many times to meet the manager of the factory, we made a written complaint to the DOSTE inspection division. . . . If complaints to the factory have no effect, we complain to other authorities. In the future we will try to pressure other organizations to force Lam Thao to change, like the press, TV and other organizations. . . . The community meets one time per month to discuss pollution. I lead the meetings. The community doesn't want to get money, it wants to have a good environment. We want clean water. Now the water is sour, it has a chemical taste, but we still have to drink it. All we can make is bitter tea.

On the other hand, Community B has been largely unsuccessful in motivating the factory to reduce its impact. The chairman of the local People's Committee explained (Mr. Dien – 4/18/97),

We are seriously affected by the factory, particularly by health impacts. People have had headaches for the last 10 years. Over 1,700 people in the commune out of 5,200 suffer from headaches. Crop yields are also very low. . . . Each year we get 15 million dong (about $1,250) in compensation paid in fertilizer. We don't want this compensation. We would prefer if Lam Thao was moved.

We did not realize how serious the health impacts were until the disease rates became high and some environmental organizations investigated. . . .

People come directly to the commune People's Committee to make a verbal complaint. The People's Committee makes a written complaint to a higher authority. We only complain through the district People's Committee. We don't know what we will do in the future. . . . If the Environmental law had not been passed, we would not have received any compensation.

Another member of the community asserted (Mr. Manh—4/18/97),

We have complained many times to the commune People's Committee, but only through verbal complaints so far, which is not very effective. We sometimes discuss pollution during general meetings, but we have never had a separate meeting about pollution and health problems.

Community B suffers under some of the worst pollution impacts I have witnessed in Vietnam, including damage to peoples' health, crops, and quality of life, but the community has been able to do little to change the situation. They have never met with the factory, never written letters to provincial authorities, never written to the media, and of course, never protested. They basically follow the local People's Committee's decisions. This unwillingness, and really inability, to influence higher level decision makers is based largely on the communities' lack of external connections and lack of sense of power.

Much of Community A's success has resulted from knowing who to complain to and using their ties to local government officials to demand compensation from the factory. The community's effectiveness at winning compensations for damaged crops and water supplies has motivated Lam Thao to invest in solutions that are cheaper than compensation. The community has documented impacts from the air and water pollution, made it available to government officials and local media, and demanded official monitoring of pollution levels and impacts. Community B on other hand, has waited for the government to analyze the problems and assign fair compensation.

Conflicts within the state also influence the Lam Thao case. Two ministries are responsible for promotion and regulation of the factory: VINACHEM, the Vietnam Chemical Corporation, a division of the Ministry of Industry, owns and operates the factory; while the Ministry of Science, Technology, and Environment (MOSTE) is responsible for regulating it. MOSTE's lack of autonomy vis-à-vis the Ministry of Industry seriously hampers enforcement: the Ministry of Industry is strong enough to effectively veto environmental regulation that MOSTE attempts to advance. Community members have thus focused their complaints on the provincial People's Committee and the DOSTE, demanding local protections from centrally controlled industry.

The Lam Thao case demonstrates that community cohesiveness is by no means enough to win environmental improvements, particularly with regard to state enterprises. In this case, the community that is more cohesive but lacks linkages ends up gaining very little in its struggle for environmental improvements. Community B is cohesive but disconnected, and unlikely to win anything more from Lam Thao than a few bags of fertilizer, an ironic compensation for the pollution it faces. When Community A combines internal cohesion and external linkages (in the form of connections to higher authorities) it is able to pressure the state and firm to take action.

3.2.4 Owner as Regulator: Ba Nhat Chemicals

When Mr. Tien left his window open, as people without air-conditioning are forced to do on hot summer days in Hanoi, within an hour much of his one-room apartment would be coated with a fine layer of white powder. For Mr. Tien, a retiree from the government, living next to the Ba Nhat Chemical factory meant living with calcium carbonate dust, the noise of grinding rocks at all hours of the day and night, and the respiratory problems that haunted the neighborhood.

Ba Nhat has been producing chemicals in this area since the 1960s, when three small cooperatives were merged into a city-owned company operated by the Hanoi Department of Industry. Output at the small factory, which employs 200 people, had grown over the years, as had the pollution that rained down on the apartment buildings just five meters from Ba Nhat's walls. Figure 3.5 highlights how close residential apartments are situated with respect to Ba Nhat chemical factory.

Pollution had been serious since at least 1987, when the community began complaining in earnest about the impacts of the factory's production. Over a period of twelve years, community members wrote letters to all levels of the government, including the National Assembly; submitted a letter to the courts; motivated journalists to write articles; and even wrote

Figure 3.5
Chemical manufacturing at Ba Nhat with residents living nearby

their own articles and paid to have them published. These actions were co-ordinated by the "Committee Against the Pollution of Ba Nhat," which met regularly to strategize about the factory.

The community around Ba Nhat has all of the critical traits necessary to motivate action on environmental issues. It is cohesive, has a high technical capacity, and has good connections to government officials. It is the best educated of any of the communities I studied, made up of government employees and current and retired professors from the nearby Hanoi Polytechnic University. Many people in the apartment complexes next to the factory have been living there since they were built. Several older people, including retired professors, participate in the Committee Against the Pollution of Ba Nhat, the first "NIMBY-like" organization in Vietnam that I know of. Residents are relatively well-off, solidly upper-middle-class by Vietnamese standards. The community is in an urban area, close to the halls of power. It even has access to a wealth of damning environmental data.

The community around Ba Nhat is not as unified as rural villages in Vietnam, but it has nonetheless been able to coordinate actions and strategies over time. And although the community is largely respectful of authorities and is sure to follow official procedures, people have been willing to publicly challenge the state. The community also has excellent connections to different levels of government, partly because of individual community members' status and positions, and to the media.

Despite all of these positive attributes, the demands of the community to clean up or move Ba Nhat were blocked for many years. At different times, the community's goals were blocked by Hanoi city politics (because the Department of Industry was stronger than the Department of Science, Technology, and Environment), by lack of capital, by state rules on selling land, by other communities and their NIMBY concerns, and by a belief that it is not worth spending money on old, dying factories. However, through sustained community pressure they were able finally to overcome all of these blockages. Without this pressure, the factory would have likely remained in place until it collapsed from old age.

Community members were first successful in 1990 in pressuring the government to commission a study on the factory's pollution. The results found that 3,000 people were adversely affected by the pollution, which included: carbon monoxide emissions seventy times the legal limit, dust ten times, sulfur dioxide four times, and other toxic gases five to seven times over the permitted levels (Nguyen 1996). By the early 1990s, virtually everyone seemed in agreement that the factory was a problem. Every level of government imaginable had been contacted. Data clearly showed the factory in violation of environmental laws. Nonetheless, the factory continued with business as usual.

In response to this inaction, community members in coordination with the *Phuong* People's Committee filed a lawsuit against Ba Nhat. For almost a year, the city bureaucracy passed the buck from agency to agency, essentially killing the suit, but also agreeing that they would consider relocating the factory. Another year passed (now 1993) before Ba Nhat announced that they had found a site on which to relocate.

Unfortunately in 1994, after the factory's move had been approved by government officials, the community to which Ba Nhat was trying to move began fighting the relocation. As Dr. Khien, the head of the Hanoi DOSTE explained, the "decision of the new location in Mai Dong . . . met with strong opposition from local institutions, community and companies. As a result, Ba Nhat could not move to this place though they had spent money for a feasibility study" (Khien 1996). In 1995, the head architect recommended the factory move to a new site in Thanh Tri district. The Hanoi DOSTE supported this move. However, once again resistance by the receiving community blocked the move.

One community leader explained the history of community actions (Mr. Cu—4/19/97),

In 1982 (when I moved here) the area was already polluted, but we started to write letters in 1987. People met and discussed the problems and wrote letters. We've written more than 100 letters. We've written to the city government, to the state government, to the Ministry of Health, and to the DOSTE. . . . We wrote articles ourselves and paid to have them printed, and then journalists got interested. Newspapers play a very important role, combined with complaints from people.

We created a "Committee Against the Pollution of Ba Nhat," which belongs to the *Phuong* People's Committee. We don't have regular meetings, but if we have a specific problem we call a meeting. There are ten people on the committee.

In 1995, the Hanoi PC agreed that the factory was polluting and that it should be moved. We had three requests in 1995: stop operation; pay compensation for health impacts; and move the factory outside the city. We don't care much about compensation, but we use it as a tool to get the factory to stop producing. We say they have to pay compensation or stop operating.

DOSTE tells us the reason the factory cannot stop is (1) the government needs the taxes from Ba Nhat, which is only 10–20 million dong per year; (2) the factory employs 200 people that need the jobs; and (3) the factory supplies important materials to other state industries.

The community says "is money more important than health?" Are 200 jobs more important than 3,000 peoples' health? The government thinks about money more than people.

The factory has always said, "We are moving anyway, so there is no need to improve conditions." Because the government says they have agreed to change and to move the factory, we can't really argue with them. They claim they have listened to us.

Another community leader discussed the role of outside actors (Mr. Tien—5/28/97),

The media has been an important factor, showing everyone in Hanoi and the country about the problem. This has helped to accelerate the process. . . . There has been a positive effect from the TV and newspapers. We have heard many times that the factory would be moved—for over three years in fact—but nothing has happened.

The *Phuong* People's Committee supports the community. The district People's Committee supports the community. But at the city level, they always answer that they will solve the problem, but their enforcement is not strict. The Environmental Management Division is not powerful enough.

Even facing these long and drawn out frustrations, community members continued to push for the factory's relocation. When one strategy did not work, the community turned to another. Finally, in 1998, after more than ten years of community complaints, the Hanoi government announced that it would physically move the factory out of the city center to a rural area with an existing chemical complex. As one community member explained (Ms. Anh—2/27/98), "The people are very happy about the decision to close the factory. They have been fighting for a long, long time. They now realize that community pressure was effective."

City government agencies are at the center of the Ba Nhat decision-making process: the Hanoi Department of Industry (DOI) owns and manages the factory and the Hanoi DOSTE is responsible for regulating it (although community members complain that responsibility for environmental management of the factory is not well defined). Both agencies report directly to the Hanoi People's Committee. Within this political system, the DOSTE is much weaker than the DOI. In fact, until the Ba Nhat case, the DOSTE had never been able to shut down or move any of the DOI's 200 factories, despite repeated promises to do so.

What the community needed and failed for years to find was any leverage over the Department of Industry. For state-owned enterprises such as Ba Nhat, environmental reforms necessarily involve one state agency pressuring another to make a change. As the National Environment Agency does not have jurisdiction over city-owned factories, this case boiled down to a political battle between the promoters and regulators of Ba Nhat within the Hanoi city government. Failing to motivate changes in the Hanoi bureaucracy, community members took their complaints to higher levels, petitioning the National Assembly and even the prime minister.

Community members have taken a wide range of actions over the last twelve years. As mentioned, the community wrote its first complaint letter about Ba Nhat in May 1987. In response, the city government investigated and determined that the factory's pollution was not toxic. The community disputed this finding for two years, sent letters to the highest government authorities without response, and then finally convened a meeting of scientists and journalists to discuss the issue. Representatives of the People's

Committee were present at this meeting and promised action. However, debates between community members and the Hanoi People's Committee continued for several years without any action taken against the factory. The community nonetheless continued to send letters to government agencies.

DOSTE staff explained in interviews that they had faced a series of battles to win this decision. Continued (and escalated) community pressures were critical to strengthening the position of the DOSTE and, I believe, ultimately tipped the scales towards moving the factory. As one government official explained, "Pollution was the key issue on motivating the move. There were many complaints from the public, and the National Assembly representative from Hai Ba Trung [Ba Nhat's district] worked to push forward the decision. Ba Nhat is the first factory in Hanoi to be moved by force because of public pressure" (personal interview—12/26/98).

Clearly community pressure was critical in this case. Ironically, one of the most significant barriers to moving Ba Nhat has been other communities blocking its relocation to their neighborhoods. It seems in Vietnam—as in other countries—it is easier to block new projects than to promote a change in an existing factory. But over time the community has been able to overcome even these impediments. As Khien (1996) asserted, "Without persistence, patience, and efforts of people in the surrounding area, the matter could have been forgotten easily." The community has played a critical role in keeping pressure on the government, applying this pressure to multiple levels of decision makers, using the media to extend their message, and even using scientific analysis to marshal its own evidence on the pollution. In the end, this cohesive, connected, strategic, and persistent community did finally win.

Once again, the Ba Nhat case shows that community capacity and cohesion alone are not enough, and that linkages are critical but complicated. This is by no means an isolated community, but its connections with the state were frustrated by other powerful interests for more than a decade. With little autonomy, the Hanoi DOSTE is almost powerless to regulate polluting state enterprises that provide jobs and tax revenues to the Department of Industry. Only connections that were able to invoke power above the Hanoi People's Committee, and extensive public pressure were enough to overcome the dominant position of the DOI.

3.2.5 Growing Up with Pollution: Environmental Change at Viet Tri Chemicals

Viet Tri City rises up out of the Red River delta as a testament to Vietnamese industrial development. The town was created in 1963 as a base for industry away from the wartime target of Hanoi. Virtually everything had to be imported to make Viet Tri into an industrial center: technicians from China and Russia designed and built the factories; workers came from around the Red River delta to begin the process of proletarianization; raw materials were floated down to the confluence of the Lo and Red Rivers where a town was quickly being constructed. Viet Tri is an example of Vietnam's development planning—designed to expand industry into rural areas while maintaining central management.

In the rush to create an industrial complex in Viet Tri—which came to include a paper mill, a sugar mill, a beer factory, a particle board factory, a nearby fertilizer plant, and the Viet Tri Chemical factory—little attention was paid to the environmental impacts of industrial activities. Soon, Viet Tri became known as the most polluted city in Vietnam, and stories of the dusty, dirty town spread throughout the country. By the end of the 1980s, Viet Tri's factories had gained well-founded reputations for being major polluters, and Viet Tri Chemicals became known as one of the most egregious polluters in an increasingly "unlivable" city. In 1988, city authorities went so far as to petition for Viet Tri Chemicals to be shut down.

Give this history, it is surprising to learn that by 1993, Viet Tri Chemicals had been "recognized as having made the most positive contributions to [a] cleaner environment" of any chemical plant in Vietnam (Nguyen 1993). During tough economic times, the factory management succeeded in securing a loan from the central government that allowed them to significantly upgrade their sodium hydroxide production equipment, plant 230,000 trees near the factory, institute an emissions monitoring program, and train their workers in better environmental practices. Although the factory had by no means solved all its environmental problems, it significantly reduced toxic emissions in roughly four years.

Asked why these actions were taken, the vice-director of the factory explained simply that the changes were necessitated "by community pressure." This is likely an oversimplification, but it does highlight the importance of community demands in the transformation of a seemingly insulated state enterprise. The factory justified its capital investments in equipment changes by pointing to both economic and environmental benefits. Switching from graphite to titanium electrodes in its electrolysis process helped the factory substantially reduce its energy costs, thereby lowering overall production costs. At the same time, the change reduced lead emissions and accidental releases of chlorine gas. Later, after a round of particularly vocal community complaints, the company established an emissions monitoring program.

The community around Viet Tri Chemicals is in general a middle- and working-class urban community. They are, however, quite diverse. Some people work in the chemical, paper, beer, and other neighboring factories, some work for the Viet Tri city government, some are petty traders, and others raise fish and farm rice nearby. Despite this diversity, and a lack of obvious connections, the community has been able to develop good ties to local government authorities and to exert considerable pressure on the chemical factory.

Because Viet Tri was a planned industrial community constructed in the 1960s, most of the community members moved to the town from other areas. Viet Tri is thus a relatively young community, which makes it somewhat unique in Vietnam. Nonetheless, the community around Viet Tri Chemicals has been able to mobilize its members to write letters and even stage "gatherings" in front of the factory gates.

Living with pollution over the last forty years, the community around Viet Tri has developed a strong awareness of environmental issues. As one reporter noted, "The industrial city of Viet Tri . . . is acknowledged to be so polluted that state employees are paid a 12 percent premium to compensate for the risks of living there" (Stier 1996). Half of the workers from the factory live nearby. The acting director lives only 200 meters from the factory gates. Community members, including some who work in the plant, have written numerous complaint letters about the pollution problems to the factory managers and local People's Committee. The community around Viet Tri Chemicals thus has a very high level of awareness, technical knowledge, and capacity for responding to environmental problems.

Workers were intimately aware of the health impacts of the factory's pollution due to a rash of chlorine gas releases in the 1970s and 1980s.

Groundwater pollution was also hard to ignore, as it contaminated the wells of many of the workers who lived nearby. The community around Viet Tri, although including groups with very different interests—workers, managers, farmers, and other residents—is fairly unified on pollution issues. Awareness, and close connections to the impacts of emissions, have led the community to exert significant pressure on the factory over the years. In 1996, when a late-night spill from the factory's detergent line killed fish in a cooperative fish pond, community members were at the factory gates the next morning with evidence and demands for compensation. Several government officials mentioned these spontaneous "gatherings in front of the factory" as a major pressure on Viet Tri Chemicals.

The factory has felt the pressure of these community complaints for many years, as the vice-director explained (Mr. Tuyen—6/24/97),

One time the chairman of the Viet Tri city People's Committee sent a formal letter threatening to shut down the factory. This was based on community complaints.

Even before that, we recognized the environmental issues in 1986. We have taken a number of actions. We have planted 230,000 trees since 1986. We reinforced our safety regulations. We have increased the environmental awareness of our workers. And we changed our electrodes.

We have a system of monitoring stations at the weak points in the production line. . . . We set it up in 1990 because of complaints from surrounding community members.

Now we have a program to allow community members into the factory. They can be on teams made up mainly of officials from the *Phuong*. They come once or twice per year. The concern of the community is whether the production is causing environmental problems. During the visits we explain the real situation to them. They can tour the production lines.

Every factory wants to have a good reputation with the community around them. After the fish kill caused by our detergent factory we dug a pond, or really a settling tank, to contain the wastewater. We also had to pay about 5 million dong in compensation to the villagers. We built the settling tank because of the complaints from the community members and because we don't want to pay compensation in the future.

A future incentive for us comes from community pressures. We might be threatened to be shut down and we might be fined by environmental inspectors.

The vice-chief of police of Minh Nong commune asserted that pressure on Viet Tri Chemicals increased in the 1990s (Mr. Khoa—5/28/97), "Complaints have been more frequent after the environmental law. I think the law is a very good thing. If the community had not complained, the factory would not have changed."

One local official explained the formal process for community participation and advancing complaints (Mr. Loc—5/28/97),

Complaints from the community come out during meetings of voters. So then the
candidates for the election take the complaints to the People's Council. The voter
meeting takes place twice per year. We also sometimes have unplanned meetings
when a National Assembly member comes to town. Based on complaints from the
community, we prepare complaints to the upper levels—to the city People's Committee—and then the city People's Committee complains to the province.

We have a close relationship with the factory. We organize tours for neighbors
into the plant. Many people go on tours of the plant and notice the performance
of the factory has improved recently. Once per year we organize the tours. We
started the tours in 1995. There was a new election so the people running for office came up with the idea of tours—of doing something for the community. The
Phuong People's Committee implements the idea. We also meet with the factory
very often.

Now the factory has a better relationship with the community. Many workers in
Viet Tri Chemicals live in this community.

All the factories in this area have made changes on environmental issues, but Viet
Tri Chemicals has been the most active, because people thought Viet Tri Chemicals
was the main polluter in the area so the most pressure was on them.

This system of public tours of the factory is unique as far as I know in
Vietnam, and interestingly was prompted by community complaints at a
meeting of candidates for local elected office.[2] The candidates turned the
idea of touring the factory into a campaign promise that has since been
honored once or twice per year. The factory has also set up a program to
reward workers for coming up with ideas to reduce waste or improve environmental performance. In 1996, the factory awarded workers a total of
50 million dong (approximately $3,600) for ideas that were implemented.

Community interactions with the factory are not always so amiable. The
vice-chairman of *Phuong* People's Committee described more adversarial
processes (Mr. Tho—5/28/97),

People gathered in front of the factory to complain in 1995. Some peoples' homes
and metal roofs were corroded by air pollution from the factory and some animals
were affected. The smell of chemicals from the wastewater was also very strong.

That gathering was spontaneous because people thought they were seriously affected. The gathering happened after a series of formal complaints received no response. The people complained for a long time and got no response.

The community has a high awareness of environmental issues and they think if
they write a formal complaint letter it will help. They sometimes write formal letters and have community members sign them and then send them to the ward, city,
and province People's Committees.

A number of specific incidents have fueled the long-term pressure on Viet Tri Chemicals. One villager who suffered economic losses from the factory's pollution when the fish in his pond were killed, explained (Mr. Dan—6/26/97),

On the morning after the pollution spill we met with the factory. We asked people from the factory to come to the fish pond to see the dead fish. The cooperative also came to evaluate. We did not invite the Center for Environmental Management; the factory invited the Center to inspect. The factory wanted to preempt the investigation. If the sufferer invites the Center, the case becomes more serious for the factory.

We estimated the loss from the pollution was 60 million dong. But the factory only agreed to pay 5.8 million dong. The compensation value was based solely on the dead fish we could collect and weigh. The factory had many arguments against us.

Figure 3.6 shows some of the fish killed by a pollution release from Viet Tri Chemicals.

There are also conflicts within the government regarding the factory, which is a centrally managed state enterprise. The profits from Viet Tri (if there are any) go to Hanoi, while the problems stay in the community. The few benefits from the factory accruing to local officials seem to be counterbalanced by community complaints, and the fact that the factory continues to tarnish the city's reputation. It is thus somewhat understandable that local officials would support the factory's closure or at least stricter regulation.

The very real threat of being shut down, driven by both local calls for the factory's closure and poor economic performance, has led the management to take community demands seriously. One manager explained that the factory's bad reputation was one reason they were not able to access government loans. Viet Tri Chemicals thus realized that they could not respond to pollution complaints by simply building taller smokestacks or glossing over problems. Instead, the factory chose to upgrade its production methods, which helped change the company's environmental reputation, and in turn helped the factory access loans, retain workers, and sell products.

As in other parts of Vietnam, community members in Viet Tri submit complaints to the local People's Committee, which then forwards the complaints to the responsible agencies. However, two differences seem to be at work in this case. First, when community members complain about Viet

Figure 3.6
Fish allegedly killed by pollution from Viet Tri Chemicals

Tri Chemicals, it is the factory that has the burden of proof to show it is not guilty due to long-standing awareness about pollution from the factory. This is the opposite of virtually every other case I examined in Vietnam. For example, in the case of the detergent spill that killed local fish, it was the factory rather than the community that requested the DOSTE inspect the situation. The factory felt it needed data to show it was innocent. Most factories would be presumed innocent until proven guilty. Viet Tri Chemicals seems to be the opposite. Second, while Viet Tri Chemicals is a state enterprise, local government officials have made clear that they are willing to challenge the central government to resolve problems at the factory. Local elected officials appear to consider this an issue they cannot ignore. This cleavage between state agencies provides a political opening for community demands.

Viet Tri Chemicals shows that SOEs can be regulated under certain circumstances. A connected community with strong capacities was able, over a number of years, to put the factory on the defensive. Even though the community surrounding Viet Tri is diverse in many ways with varied interests and actors, it was able to mobilize effectively. This is due to both the severity of the environmental problems, the alignment of local interests—including of workers—to clean up the factory, and the ability of the community to find allies within the Vietnamese government. Through changes in state concerns and conflicts between local and central agencies, a previously insulated factory became vulnerable to community complaints. Community members were successful in establishing an effective system of alarms, and vulnerable factory managers then turned these pollution concerns into pollution prevention strategies that had both environmental and economic benefits.

Although individual disputes are not always settled to the villagers' satisfaction, community members say they can see gradual improvements in the factory. One person who lives near the factory asserted (Mrs. Nguyen—6/26/97) "I think that the environment has positively changed in the last few years. The factory now follows regulations and is afraid to be fined." Figure 3.7 shows community members who organized to demand compensation from Viet Tri Chemicals.

3.3 Understanding Mobilizations in Vietnam

Both classic studies of social movements and more recent analyses of "New Social Movements" (including environmental struggles), seek to explain motivations and processes of community mobilization and collective action. McAdam, McCarthy, and Zald (1996: 2) for instance argue that although the literature is quite diverse, "Increasingly one finds movement scholars from various countries and nominally representing different theoretical traditions emphasizing the importance of the same three broad sets of factors in analyzing the emergence and development of social movements/revolutions . . . (1) the structure of political opportunities and constraints confronting the movement; (2) the forms of organization (informal as well as formal), available to insurgents; and (3) the collective processes of interpretation, attribution, and social construction that mediate between opportunity and action." Essentially, communities take action when crises occur or political opportunities arise, they mobilize through existing networks and organizations, and they frame their actions in order to win support from other community members and the public at large.

Figure 3.7
Community members who demanded compensation for fish killed by a pollution spill from Viet Tri Chemicals

An interesting body of work has been emerging regarding these types of actions within civil society in Vietnam. This research is challenging traditional notions of communist state–society relations. Kerkvliet (1995: 415) for instance, surveys the existing debate around Vietnamese state–village relations, grouping theorists into three camps: (1) those who believe the state is all dominating; (2) those who assert the state controls civil society through various means; and, (3) scholars who believe social resistance is central to state–village relations. He concludes that the third view is more and more accepted, as "a prominent theme in the recent history of rural Vietnam is debate, bargaining, and interaction between the state and villages." Thrift and Forbes (1986) arrive at similar conclusions for processes and controversies over urbanization, showing how communities play a major role in forcing changes in state policy. Fforde (1993), Beresford (1988b), Luong (1992), and Fforde and de Vylder (1996) discuss equally surprising forms of community resistance and state concessions for a range of economic and political issues.

Collective action, of course, does not spring from the ether, but is based in existing institutions and networks—or "mobilizing structures."[3] In Vietnam these mobilizing structures include state-controlled institutions such as People's Committees, nonstate commune and village relations, churches, kinship relations, NGOs, and mass organizations (such as the Women's Union, Youth Union, and Peasant's Union). Although "civil society" appears poorly developed in Vietnam, long-standing social and kinship relations (which have weathered wars and major political changes), and sophisticated decentralized social networks strengthened during wartime, serve as the basis for thick formal and informal social institutions. These networks and organizations are critical to collective action in Vietnam.

The existence or development of a "sense of community" can support participation in local organizations and political processes more broadly. As the "social capital" literature attests, groups with high levels of trust, norms of reciprocity, effective communication, and civic networks are much more likely to take action than heterogeneous groups with low levels of trust and cooperation. Local movements often depend on these existing norms and networks to support the transition from individual complaints to coordinated actions (see for instance, Woolcock 1998, Woolcock and Narayan 1999, Putnam 2000). Coordinated actions and community participation in turn can engender increased awareness of local problems, shared needs, injustices, and pollution controversies (as in these cases). It can also lead to an increasing sense of collective power.

The Communist Party has of course attempted to control local politics. However, it has at the same time been responsive to some local demands. The state has repressed certain individuals and organizations, but it has not been able to quell all community action. Even at the height of the "planned economy," vibrant markets existed beyond the reach of the state. As Kerkvliet (1995: 414) notes, "the State is sufficiently 'mass regarding' to be at least worried about the peasantry and responsive to persistent resistance." This concern for stability, combined with the Communist Party's own ideology and rhetoric, have been used as resources that allow local community actors to influence state action.

Local political structures also influence the nature of community actions. If a political system is closed to citizen participation, local energies are often directed away from official political channels towards strategies that disrupt business-as-usual (Tarrow 1996: 44). McAdam et al. (1996: 10) argue further that "insurgents can be expected to mobilize in response to and in a manner consistent with the very specific changes that grant them more leverage." If a political system is relatively open, individuals may devote more energy to "working within the system" such as through political parties and lobbying incumbent officials. Public perceptions of political openness are critical to social responses.

The dynamics of local political authorities and elites also influence processes of community mobilization. If local authorities are perceived to be controlled by external forces, community members are likely to mobilize in different directions than if they believe decision making is controlled at the local level. For instance, community members may seek extralocal allies or target higher-level authorities that can force open the political process. Also if community members have connections to local elites, such as members of the Communist Party or People's Committees, or can find cleavages in elite alliances, they are likely to mobilize in different ways. As Tarrow (1994: 17) explains, successful movements "draw upon external resources—opportunities, conventions, understandings and social networks—to coordinate and sustain collective action. When they succeed, even resource-poor actors can mount and sustain collective action against powerful opponents."

Successful movements also "frame" their issues strategically, giving environmental issues shared meanings and definitions in order to provide motivations and justifications for action (Zald 1996). As a number of analysts have pointed out, people need to feel both that a problem is "real" and that it can be collectively solved. Framing helps focus discontent, identify targets for blame, and create connections between individual problems and community members (Gamson 1990).

3.3.1 Political Opportunities

Broad political changes in Vietnam have opened up new space for debate about government decisions and more specifically about the impacts of development. At the highest levels, even the National Assembly is beginning to look less and less like a "rubber stamp" body, with real debates occasionally breaking out, and representatives seeming to take their responsibilities more seriously.[4] The country's exposure to the global economy and to global culture and politics is also having an effect on the society. The fact that people in Vietnam can now watch CNN or log onto the Internet is making social and information control much more difficult. Global opening has also brought with it an influx of international NGOs, academics and reporters.

Another major political change involves state strategies regarding the economy. With the transition to "market-oriented socialism," the state is assuming new roles in the economy, exercising less control over direct production and more control over macroeconomic processes. One of the impacts of these changes is the state's decreased willingness to subsidize and protect state-owned enterprises. Some firms are being "equitized"—Vietnam's version of privatization—while others are being forced to compete on their own. These changes have opened up space for critical assessments of the costs and benefits of industrial activities and specific state enterprises. Community members are more and more able to question the tradeoffs between state subsidies, jobs, and pollution.

The entrance of foreign multinational corporations (MNCs) into the Vietnamese economy has also opened up space for criticizing industry. Industrialization is no longer considered the panacea for all problems, and can be discussed separately from national development. The public is well aware that some firms exploit workers and pollute the environment. The contradictions between socialist ideology and the lax regulation of foreign MNCs has opened a political debate in the country.

On the microlevel, new policies and regulations have been promulgated over the last five years that have created political opportunities for community actions around the environment. The Law on Environmental Protection (LEP), the creation of the National Environment Agency (NEA) and the provincial Departments of Science, Technology, and Environment (DOSTEs), and a decree on environmental complaints have provided state legitimacy to community concerns about pollution. The passage of the LEP has given community members a justification for their actions, and a belief that the state will respond to complaints. The LEP also seems to have opened up political space for more general community complaints. As one government official noted "before 1994 there were few complaints," but after the law was passed, the floodgates opened. The chairman of a local People's Committee in Dong Nai explained (Mr. Viet—4/10/97), "There was a very positive effect from the passage of the Law on Environmental Protection. It helped to educate people. People became more active after the law. They realized they had a right to complain." Community members stated over and over "The environmental law motivated us to write more letters" (Mr. Doanh—4/10/97). Political openings are critical for effective mobilizations.

The creation of the provincial DOSTEs have not only served to establish local level enforcement mechanisms, but perhaps more importantly, have also provided a target for community complaints. As a villager near Lam Thao explained, "We recognized the environmental problem a long time ago, but there was no organization working on environment, so we just complained to the Tam Thanh district People's Committee. They did not take it seriously. After CEM [the local environmental agency] was established we could complain to them." (Mr. Duoc—5/27/97).

3.3.2 Motivations

As numerous studies have pointed out, individuals and groups are motivated to take action regarding environmental problems for a wide range of reasons (Szasz 1994, Taylor 1995, McAdam, McCarthy, and Zald 1996). Specific insults, broader social contradictions, events that call into question the legitimacy of leaders or decisions, and changes in public perceptions, all can motivate actions. In these case studies, the main motivations for community action centered around specific grievances (focused on economic and health effects of pollution), fears of long-term impacts, and a sense of moral or political injustice.

Economic costs of pollution are fairly straightforward. Crops are destroyed, yields are reduced, fish are killed, well water is contaminated so drinking water has to be purchased. Many of the community members I interviewed could list the specific economic damages attributable to a factory's pollution. For instance, villagers near Lam Thao went into great detail about their plots of rice destroyed and yields of fruit trees reduced, and villagers near Viet Tri counted the dead fish killed by wastewater from the factory.

Health impacts, and particularly impacts on children, are the other major concern of communities in Vietnam. Although rarely compensated, health effects are in many ways considered a more legitimate reason to complain than economic impacts. However, they are also much harder to prove as we know from debates about illness clusters in the United States and Europe (Brown 1992). My interviews with community members often included long and detailed descriptions of health problems people felt were caused by a factory's pollution.

Not all types of pollution motivate community action. Pollution that community members can see, smell, or feel merits the most attention. Air pollution (which in many ways is a more democratic form of pollution) generated stronger responses from community members than did water pollution. Pollution with serious acute health impacts motivates community action more readily than pollution with chronic impacts. Community members are simply less able to evaluate long-term health impacts. People rarely had information about the toxicology of local pollutants, but they nonetheless had theories on what the chemicals might be doing to them and their children. Ecological impacts were rarely mentioned as a motivation for action unless they resulted in financial costs (e.g., crop loss, fish kills, etc.).

Some community members were also motivated to act because they felt their rights had been abused. As one community member asserted, "We want to feel that the government cares about us. The government hasn't shown much care so far" (Mr. Duoc—5/27/97). Community members clearly assess the costs and benefits of their local factories, and more specifically who is receiving those benefits, and who is paying the costs. If community members feel they are receiving few benefits from the pollution (for instance when workers are shipped in from outside the community to Dona Bochang), but are paying disproportionate costs, they are more likely to mobilize. In situations where benefits accrue to some community members (such as workers and waste recyclers at Tan Mai) community calculations of costs and benefits change, and mobilizations are much more difficult.

Some community members also appear to be angered by contradictions between socialist ideology and current realities of market economics. It is hard to determine if people believe the rhetoric of the Communist Party, or if they simply use it against the government for strategic purposes. However, whichever it is, numerous community members quoted Communist Party policies and promises when explaining their demands.

3.3.3 Mobilizing Structures

Once community members are motivated to act, they mobilize through both formal and informal structures. In only one of the cases—Ba Nhat— did community members actually create a new organization to support their cause. In the other cases, people worked through existing local networks and organizations to advance their complaints. Communities in Vietnam organize themselves both within, and outside of, official structures. Whereas many women, for instance, are members of the Communist Party-affiliated Women's Union, informal associations of women and NGO supported women's groups are also becoming more common.

At least in rural areas, membership in the hamlet remains of utmost importance. In urban areas, except for the newly mobile rich, most people also have long ties to their street or neighborhood. Most people thus look to their local People's Committee for solutions to environmental and other problems, and submit their complaints either to the hamlet or ward People's Committee. These People's Committees obviously vary. Some are supportive and open, while others are closed and corrupt. The ward People's Committee near Dona Bochang is very different than the People's Committee near Ba Nhat in Hanoi.

3.3.4 Framing the Debate

Community members around Dona Bochang recognize that complaining about dust and soot falling on their houses will win them little sympathy or action from the government or the managers of the factory. So, as the factory managers continue to argue that the pollution is harmless, the community has worked to reframe the problem as a health, accountability, and moral issue. More than just working to solve immediate pollution problems, community members are actually changing the discourse in Vietnam about industrial development and sustainability.

The first step in framing the debate has involved documenting the pollution. Community members have taken photographs of the black smoke wafting out of the smokestack and settling as a fine black layer of dust across their yards and houses. One family made a video of the pollution that they provided to the media. Community members have also documented the health problems in the community they believe are caused by exposure to pollutants. This process of documenting the pollution and writing joint letters to the government has served to raise the community's awareness about the pollution and to legitimate individual concerns. Community members have come to a shared analysis of the problem and developed a set of goals for reducing the pollution. Documenting pollution trends over time has also raised awareness of the long-term impacts of industrial development.

The community has also tried to frame the pollution as a failure of the state to meet its responsibilities. In interviews, community members repeatedly criticized the government for not enforcing the Law on Environmental Protection. Letters to the government and the media refer to the state's responsibility to enforce the law and to protect people from adverse impacts of industrial development. As discussed in chapter 6, media reports raise these accountability and responsibility issues, questioning state agency enforcement practices.

Some community members also frame pollution as a moral issue, decrying the fact that the search for profits is outweighing all other considerations. One villager summed up a common concern, "The factory doesn't care about pollution because they only think about profits" (Mr. Thanh—6/7/97). Another argued that the state had lost perspective on the purpose of industrialization. "Factories are built to serve the people, not to harm people. If the factory is hurting people it is not meeting its purpose" (Mrs. Hao—4/10/97). Community members also frame the pollution to advance more general rights to a clean and healthy environment. By connecting individual cases of pollution to broader patterns of ecological and environmental degradation, community members challenge the trade-offs inherent in the drive for modernization and industrialization.

3.3.5 Actions
Community mobilizations vary in form and intensity. Some communities follow official complaint procedures and simply write letters to government offices. Others take more militant (and sometimes illegal) actions such as protesting in front of the offending factory. Some community

members are clearly reluctant to act outside the bounds of official procedures, yet others show no fear whatsoever. For instance, one community
member at Ba Nhat stated that "Protesting in front of the factory is illegal.
The only legal way is to write complaint letters to the responsible agencies" (Mr. Tien—5/28/97). And while this community has tried many
other tactics, it has never moved to the level of protest. There are a number of possible explanations for this restraint. It may be a class issue, that
is, it is too risky for middle- and upper-class people to protest as they have
more to lose (in some ways) than poor rural people. It may also be that
peasants feel they have more right to make demands on the state, whereas
urban people are often state employees.

At the same time, residents around Viet Tri explained that when letters
have failed to generate state action, community members have taken their
complaints to the factory gates. These "gatherings" in front of the factory
made both the factory's managers and the local government officials very
nervous. Other community members have reported actions that sound surprisingly similar to civil disobedience. For instance, two men living near
Dona Bochang explained their plans to block the factory's wastewater
pipe if the pollution did not stop. "We are waiting for the rainy season. If
nothing has changed we will buy soil and fill in the wastewater pipe to
block it. All the families in the *khu pho* will block the sewer together. We
will let Dona Bochang know that we are unhappy" (personal interview—
4/11/97). They knew this would be illegal and would likely cause damage
to the factory, but they felt they had no other option to get the attention of
the factory and the government.

A community's level of trust in the state influences the types of actions
it is willing to take. Some communities seem to have unlimited patience
with the government despite agency failures to enforce legal responsibilities. Other communities seem highly suspicious of state agencies. Many
community members spoke of opaque procedures for environmental regulation, and of their frustration with never seeing results from government
inspections. Over and over I heard the complaint that "We never heard
back from the DOSTE." This frustration leads to the impression that the
environmental agency is closed (and/or corrupt), which motivates community members to look for other avenues of influence.

When petitioning the state fails, or seems bogged down, communities often look to the media for support and amplification of their concerns. Although the media is officially controlled by the government, people nonetheless view it as a more sympathetic institution with the power to exert pressure on state agencies. Often the second step in community actions is a letter or call to the media, either providing them with details of the pollution impacts or asking them to come out and report on the situation for themselves.

In letters to the government and media, community members make a range of requests.[5] The first request is always that the pollution, and the impacts of the pollution be stopped. People want the government to do something to stop the health impacts they are experiencing, and to protect them from future economic damage. When there are clear economic impacts, community members also request that the government force the company to pay compensation. People almost always qualify this request with reference to their higher priority for a clean and healthy environment. Community members recognize the dangerous precedent of being paid to live with pollution. Some community members even argue against compensation, saying "The community wants waste treatment not compensation" (Mr. Dung—6/2/97). Community members around Ba Nhat use the demand for compensation as a tactic to put pressure on the firm. The threat of financial costs are used to motivate the firm to clean up or relocate.

Communities in Vietnam very rarely demand that a factory be shut down. The imperative of job creation is so great, that even unprofitable firms are considered too important to close. Essentially the social costs of shutting down a factory trump all other social or environmental concerns. Even Ba Nhat, which most outside observers would say should be shut down based solely on its abysmal technical and economic performance, is targeted for relocation rather than closure.

3.4 Characteristics of Effective Communities

Community actions often fail to motivate firm actions to reduce pollution. The Tan Mai case for example, involves a community that has sporadically demanded pollution reductions, but that has failed to motivate the firm or the state to take action to reduce the pollution. However, on the other hand, Community A in Lam Thao has taken similar actions, and has been successful in motivating changes. So what makes an effective community?

Although there is no prototypical community that succeeds in pressuring a firm or the state, a number of features of successful communities stand out, including: cohesion; strong leadership; linkages between the community and local government authorities; and a sense that the pollution is unjust. Once mobilized, effective community action is strengthened by: proof of environmental harm; sustained pressure on the firm and the state (there are few easy victories); finding means of leverage over a state agency or ways to make a state agency accountable; use of outside resources to pressure the state or the firm; and continuous monitoring of firm performance and state actions. Essentially, successful cases involve communities with strong internal social ties and supportive external linkages that forcefully, strategically, and consistently pressure an agency or firm to take action.

Strong internal social ties are critical to overcoming collective action problems regarding pollution. Social cohesion is at the heart of successful community organization and action. Shared interests, shared identities, and the community's history of organization all play a role in the community's level of unity. Organization and mobilization are also influenced by the social background of community members and the ability of the community to overcome collective action problems. Vietnam's socialist past, and the system of local People's Committees, has strengthened this social cohesion. As Fforde and de Vylder (1996: 49) argue, "It is almost impossible to overestimate the importance of [the local collective] to Vietnamese society."

Table 3.1 presents some basic demographic information on the case study provinces. Ba Nhat is located in Hanoi, Dong Bochang and Tan Mai in Dong Nai province, and Lam Thao and Viet Tri in Vinh Phu province (which became Phu Tho province during the study).

As table 3.1 shows, Hanoi is the most urban and wealthiest province in the study, but also the most unequal in income distribution. Vinh Phu is the poorest province, but also the most homogenous as it is largely rural with fairly even income distributions. Dong Nai is a formerly rural province, with low educational attainment, that is fast industrializing. This is leading

Table 3.1
Community Demographics

Variable	Nation	Ha Noi	Dong Nai	Vinh Phu/ Phu Tho*	Source
Population in 2000 (in '000)	77,916.0	2,736.0	2,039.0	1,273.0*	1
% Urban in 2000	23.9	57.8	30.8	14.1*	1
Population density in 1997 Persons/km2	219.0	2,556.0	119.0	357.0*	2
Per capita income in 1996 '000 Dong	226.7	379.3	324.2	180.9	3
Lowest quintile of Income '000 Dong	78.6	89.2	78.7	61.4	3
Highest quintile of Income '000 Dong	574.7	941.5	747.2	401.5	3
Difference between lowest and highest quintiles (1000 Dong)	496.1	852.2	668.5	340.1	3
% Enrollment of children under 3 yrs in nursery school	7.0	10.1	1.9	11.4*	3
% Enrollment in upper secondary school	28.2	56.4	28.9	35.3*	3
% Households using electricity		95.6	30.7	57.3	4
% Households with running water		5.1	0.6	0.2	4
% Owning Television		44.5	32.8	24.7	4
% Owning Radio		44.1	43.9	48.4	4
% Owning Motorcycle		17.5	22.0	6.9	4

Notes: *Vinh Phu province was split into two provinces—Vinh Phuc and Phu Tho—in 1996. Data marked with an asterisk are for Phu Tho province. Data without an asterisk are for Vinh Phu province.
1. General Statistical Office. 2001. Statistical Yearbook 2000. Hanoi: Statistical Publishing House.
2. General Statistical Office. 1998. Statistical Yearbook 1997. Hanoi: Statistical Publishing House.
3. General Statistical Office. 2000. Figures on social development in 1990s in Vietnam. Hanoi: Statistical Publishing House.
4. General Statistical Office. 1996. Summary of Rural Census, 1994.

to growing gaps in wealth and income in Dong Nai, with ownership rates of motorcycles one clear indicator of new wealth.

These general demographics are important indicators of community resources, capacities, and homogeneity. However, the specific case study communities also vary in their levels of cohesion and linkages to external powers. Although these factors are much harder to measure, table 3.2 provides a general assessment of key characteristics of the communities based on the case study research, and in particular, on an analysis of their effectiveness in mobilizing for environmental improvements.

In the cases in which the communities had basic capacities, were relatively cohesive, and had linkages to the state (Dona Bochang, Viet Tri, and Ba Nhat) the community was successful in pressuring for pollution reductions. However, even with these characteristics, changes can be blocked or delayed for many years by powerful state interests. And a lack of cohesion (Tan Mai) or linkages (Lam Thao Community B) can impede a community from successfully organizing for pollution reductions.

The community around Dona Bochang appears to be much more cohesive than the other communities I studied. The community has moved beyond traditional kinship ties, or historical networks, forming a collective consciousness and cohesion. This may be explained as Woolcock (1998) asserts, by the fact that "integration" or "social capital" is high in groups with distinct cultural differences from the mainstream, those that are engaged in regular conflicts with more powerful groups, those suffering a high degree of discrimination, and those with a high degree of internal communication. The Catholic community around Dona Bochang fits most of these criteria.

Social connections of course are not all equal or beneficial. Certain connections can lead to corruption and nepotism, particularly in the case of local-level enforcement of laws. Close connections between factory managers and workers, such as at Tan Mai, can serve to block action against a firm. Ties within the Tan Mai community are not enough to overcome these connections. In order to balance potentially corrupting forms of local level ties, a number of analysts have argued that it is necessary to build connections to external sources of power. Woolcock (1998: 24) for instance asserts that "high levels of integration, can be highly beneficial to the extent they are complemented by some measure of linkage." As he

Table 3.2
Characteristics of Communities

	Capacity	Cohesion	Linkages	Outcomes
Ba Nhat	High capacity	Cohesive	Linkages	Success after many years
Dona Bochang	Medium capacity	Cohesive	Linkages	Success
Tan Mai	Medium capacity	Strong divisions	Some linkages	Failure
Lam Thao Community A	Low capacity	Some divisions	Linkages	Partial success
Lam Thao Community B	Low capacity	Cohesive	No linkages	Failure
Viet Tri	Medium capacity	Some divisions	Linkages	Success after many years

explains in regards to development dilemmas, "the initial benefits of intensive intra-community integration, such as they are, must give way over time to extensive extra-community linkages" (1998: 25).

Effective communities thus foster ties to extralocal actors. For example, community members around Dona Bochang have developed relationships with reporters, and community members around Ba Nhat have connected with members of the National Assembly. These linkages to extralocal actors are critical to amplifying community demands and increasing the community's political leverage over other actors. Connections to an agency in the state (even if it is not an environmental agency) may be critical to efforts to advance community demands within the government. For instance, community members around Viet Tri have successfully leveraged connections to the city People's Committee into pressure on the managers of Viet Tri Chemicals. As Evans (2001: 20) notes, "communities . . . do not necessarily have to 'capture the state' in order to elicit favorable responses from public institutions. Creating alliances with the specifically relevant parts of the state may be sufficient." Which means that "Mobilized communities can have an impact in particular arenas of deliberation and decision-making within the state, even if the state as a whole is focused on other agendas" (p. 21).

Individuals obviously matter in community organization and mobilization. Individual skills, willingness to take risks, charisma, and commitment all play a role in leading community members to act. The most successful communities have some form of leadership. Leaders are responsible for mobilizing other community members and for developing the strategies that determine the success or failure of a campaign. Smart leaders mobilize broad public support and build alliances with elites, workers, and the media.

3.5 Limits to Community Action

Community actions can also fail. Divisions within the community, poor organizing, ineffective strategies, and the inability to find leverage over a recalcitrant state agency, all stand in the way of effective community action. Communities can also be arenas of conflict, and otherwise incapable of mobilization. In the case of Tan Mai, the split between those who benefit from the factory and those who pay the costs of the pollution is enough to block cohesive community action.

Some firms are particularly adept at capitalizing on these divisions. Firms can pay people off, threaten a community with capital flight, or attempt to discredit the antipollution contingent with other community voices. For example, as one government official explained, "Lam Thao is very active in settling complaints. They go directly to communes and negotiate compensation. So now the most affected communities don't complain. Sometimes Lam Thao tries to bribe commune leaders. But bribery in some situations does not work if the problem is too severe" (Mr. Hung—6/27/97). Dona Bochang has also attempted to hire community activists to work for the company.

It is often difficult for community members to sustain campaigns over long periods, leading to community initiatives being defeated simply by time and attrition. Firms almost always have the resources to outlast community members. Conversely, community members sometimes give up after winning a small victory or after an immediate goal is met. Limited victories can have a demobilizing effect. For instance, it will be interesting to see if the community around Dona Bochang will be able to continue its mobilization and successfully pressure for a resolution to the water pollution problems, or whether they will be demobilized.

Communities also sometimes come up against powers they simply cannot influence. Vested interests (for instance in expanding industrial activity or protecting existing firms) often work to block environmental initiatives. Local government agencies are sometimes unresponsive to local pressures, and corruption at the local level can stymie even the best-organized efforts to pressure the state or a firm. When decisions about production and pollution are made extralocally, it is extremely difficult, and sometimes impossible, to influence these decisions at the local level.

Communities can also sometimes mobilize against each other. In the Ba Nhat case, people living in the community where the factory was to be moved complained forcefully and pressured the Hanoi People's Committee to block the move. One community's mobilization can block another's best efforts.

Even when communities organize successfully, there are limitations to what they demand and achieve. With little data and no training, community members often end up only complaining about localized, short-term, acute impacts of pollution. This type of pollution likely accounts for a significant portion of industrial pollution in Vietnam at present. Nonetheless, this focus severely limits the range of environmental issues that become priorities for state action. Furthermore, with no knowledge of technical alternatives, communities tend to push for pollution control rather than prevention simply because their main concern is stopping local emissions.

3.6 Conclusions

The ultimate concern of community members is whether or not they are successful in reducing an environmental hazard. The question of whether communities mobilize is thus only a precursor to the more important question of whether pollution is reduced.

Different forms of community mobilization result in varying state and firm responses. Communities can interact with state agencies and firms in a purely adversarial manner, they can work through state-organized and facilitated processes (some of which are effective and some of which are corrupt), or they can submit to state-dominated processes that allow little opportunity for grassroots participation. The case studies indicate that one promising mechanism for pollution reduction in Vietnam involves communities both pressuring and coaxing the state to regulate a firm. Adversarial actions are sometimes needed, as are cooperative relations.

The right types of community pressure on the state can serve multiple purposes. They can act as a check and balance on state action and inaction, they can shine a spotlight on corruption, and they can strengthen a community's position for future struggles. As Woolcock (1998: 8) argues, "a broadly participative civil society not only contributes the 'checks and balances' on government action, but provides citizens with the organizational skills and information they need to make informed decisions." And most importantly, community action can help to motivate regulation where before it was ineffective. As one community member summarized (Mr. Doanh—4/10/97), "Community complaints with the support of the government can make a factory change." The process of making a factory reduce its pollution is obviously complicated. We turn now to these firm dynamics.

4

Plans, Profits, and Pollution Decisions: Motivations, Resources, and Firm Responses

Activists and academics have fairly clear expectations of how multinational firms such as Nike respond to state and public pressures. An underlying market logic is assumed to guide profit-maximizing, brand-sensitive firms in their responses to environmental, labor, and other regulations, as well as to public concerns. And although western firms often diverge from this logic, firms in Vietnam are even harder to predict. In fact, different types of firms with varied decision-making processes and incentives, appear to respond quite differently to environmental issues, challenging existing conceptions of state–firm relations, and ultimately of how regulatory processes work on the ground. For firms in Vietnam, regulations and "market signals" are tied up with broader dynamics.

Beneath statistics on GDP growth, imports, exports, and industrial output, lies a complex story of Vietnamese industry. State policies and plans, individual investments, product changes, innovations, firm reinventions, equipment breakdowns, and financial losses, make up the messy reality of Vietnamese industrial development. In the state sector, firms that lose money year after year remain in business. Loans that should go to productive investments are diverted to seemingly doomed enterprises. Technology that should be in a museum remains in operation. And the government continues to subsidize unprofitable, polluting factories, which on the whole provide very few jobs.

Viet Tri Chemicals is but one example. In a different country, the factory would probably have been shut down long ago. The company loses money selling its main products, and the production of these unprofitable chemicals causes serious pollution. Instead of being shut down, however, Viet Tri Chemicals recently received a new lease on life. In 1998, the government extended import tariffs on foreign chemicals to protect the company's markets. Even with these protections, the company still cannot compete against imports and will likely stop producing basic chemicals in the near future. The factory's manager hopes to reinvent the company as a cosmetics manufacturer, a plan only slightly less unlikely than the strategy of another state chemical company (currently producing fertilizers) to diversify into champagne production.

The "private sector" in Vietnam is equally perplexing. Foreign multinational corporations (MNCs) form joint ventures with state-owned enterprises, blurring the lines between private and state control. MNCs that are staunch capitalists hire well-connected Communist Party members to assist in their business dealings. Saturated markets are entered based on assumptions of the long-term growth potential of the Vietnamese economy. One example is the fourteen foreign auto companies that raced to set up factories in Vietnam with a combined capacity of over 200,000 cars per year. Vietnamese consumers purchased only about 13,000 of these cars in the year 2000.

Clearly Vietnam is a country where economic analysts operate at their own peril if they assume markets function according to neoclassical theories or that actors always maximize profits. Instead, complex and sometimes contradictory dynamics influence industrial decisions during the country's transition process.

Although this study focuses on a subset of industrial development decisions, those related to environmental impacts, these decisions are sometimes even more difficult to evaluate and understand than direct production decisions. Environmental decisions are often integrated into broader production strategies and are thus difficult to separate from other pressing concerns. Firms attempt to resolve environmental issues as they struggle to compete, survive, or expand. Many firms also often shirk environmental responsibilities, bury environmental duties deep within their bureaucracies, or seek to externalize environmental issues altogether. However, when motivated, some firms do take actions to respond to environmental pressures. And when pushed far enough, some of these firms move beyond claims, promises, or taller smokestacks, to actually resolve pollution problems.

Vietnam provides an interesting context for examining how different types of firms, with different motives and internal decision-making procedures, respond to environmental concerns. Some firms are innovators, turning environmental challenges into opportunities for product and process improvements. Most, however, require significant external pressure to motivate even the smallest actions to reduce pollution. In responsive cases, at some point, due to a combination of internal and external dynamics, firms evaluate (or reevaluate) the costs and benefits of business-as-usual, and decide to make environmental improvements. Is it regulation that drives this recalculation of costs? Fines and compensations? Consumer pressures? The perception that in the long-term pollution reduction will cost less than the liabilities of pollution? Or is it simply the environmental awareness and concerns of factory managers?

This chapter reviews how the case study firms respond to pressures for pollution reduction, focusing on two main issues: (1) what motivates firms to take action; and, (2) what capacities are needed for firms to respond effectively. Motivations include both external pressures (such as market competition, state regulation, community pressures, and NGO campaigns) and internal decision-making processes. Capacities include both resources and processes, that is, both technical and managerial capabilities and the decision-making processes that allow a firm to overcome barriers to pollution reduction. The cases provide insights into the capacities needed to turn motivations into actions, and to overcome technical, organizational, and financial barriers to environmental improvements.

4.1 Firms in Vietnam

Vietnam is undergoing simultaneous economic and political transitions that directly influence industrial activities and the structure and strategies of industrial enterprises. Firms are assuming multiple forms and employing varied strategies as the nature of the market has shifted over the last several years. Even individual firms can sometimes take on multiple identities. This research examined four types of firms: (1) centrally managed state-owned enterprises (SOEs); (2) locally managed state-owned enterprises; (3) foreign-owned firms; and (4) joint-venture firms (jointly owned by the Vietnamese government and a foreign firm). These firms vary in more than just ownership. They also vary in strategies, management procedures, incentive structures, stakeholders, and responses to environmental concerns.

4.1.1 State-Owned Enterprises (SOEs)

State-owned enterprises (SOEs), a dying breed of firm around the world, represent not only the past, but also if the government has its way, the future of Vietnamese industrial development. These firms reflect state power in the economy, and are as much part of a political as an economic strategy. The Vietnamese government is interested in expanding the state sector both for employment purposes (which has a political component), and to strengthen the state's overall role in the economy. Vietnam's "socialism with a market orientation" places a high priority on state participation in productive activities.

Over the last thirty years, government bodies, primarily line ministries, made almost all decisions regarding large-scale industrial activities. Control over production was divided between state (central versus local) and nonstate (private and cooperative) enterprises. Product mixes and quantities of outputs, as well as the specifics of technological and resource inputs into production were determined by the Ministries of Industry. Capital allocations were proposed by the Ministries of Industry and approved by the Ministry of Finance. The largest SOEs, which primarily focused on heavy industry, were controlled from Hanoi. In 1992, the state sector accounted for 74.4 percent of the value of industrial output, cooperatives accounted for only 2.4 percent, and the private sector contributed 23.2 percent of industrial value. By 2000, this distribution had changed significantly as the state-owned sector's contribution to industrial output had fallen to only 46 percent. However, while the total number of state enterprises is shrinking, the enterprises that have survived are increasing their output (EIU 2000b).

The promotion of State-Owned Enterprises (SOEs) remains a central concern for the government. The Prime Minister recently announced plans to reorganize and consolidate SOEs to make them more efficient and to maintain their leading role in the economy. This is basically an extension of past policies to support SOEs through soft loans, implicit subsidies, preferential export licenses and access to land, and other support policies. The government continues to protect sectors such as cement, steel, paper manufacturing, sugar refining, and car assembly through tax and tariff policies. At the end of 1997, there were 560 centrally run, state-owned industrial enterprises employing approximately 400,000 workers.

In 1994, the government restructured the management of centrally run SOEs into General Corporations. The creation of General Corporations was designed to rationalize SOE management, to eliminate line ministry control over enterprises, and to realize economies of scale. Approximately twenty General Corporations have been created including corporations for oil (PetroVietnam), coal (VINACOAL), textiles (VINATEX), chemicals (VINACHEM), steel, cement, machinery, paper, telecommunications, and air transport. The government is now considering establishing other "economic consortia" in sectors such as electricity, telecommunications, and petroleum that would, if successful, become essentially Korean-style *chaebol* conglomerates (EIU 2000b).

Many smaller SOEs are scheduled to be restructured or "equitized" in the coming years—a process in which they are partially privatized through the creation of stock companies. The government's equitization policy was established in 1993, but it was not until 1998 that the process really took hold. Of 1,590 SOEs located in Hanoi and Ho Chi Minh City, 606 were slated to be equitized. Nationwide, the government has a target for the year 2005 of equitizing 1,300 small SOEs, selling or leasing another 500, merging 350, and dissolving or declaring bankrupt 368 firms.

The SOE structure comes directly out of past policies and Vietnam's implementation of essentially a Soviet model of industrial development and modernization. From the 1960s to the mid-1980s, government development plans focused primarily on the promotion of heavy industry, which received the overwhelming majority of state funds. Key industries included: iron and steel, chemicals and fertilizers, cement, coal, vehicle manufacture, machinery production, as well as selected light industries such as food-stuffs and textiles.

In the past, control over industry and the stated objectives of government planners motivated factory managers to focus primarily on meeting production targets. Profits, in the capitalist sense, were not a critical factor in production decisions, although accumulation of surplus was a stated objective. In SOEs, profits were separated from production decisions, so that political objectives could drive decisions on what to produce and in what quantities. Resource inputs were priced through planning decisions or considered "free" goods (such as water), resulting in a general atmosphere of inefficient use of resources and ineffective production methods. Environmental considerations were rarely included in production decisions.

As analysts have noted (Beresford and Fraser 1992: 10–12),

The traditional socialist economy has inbuilt mechanisms leading to enormous waste of resources and inhibition of technical change, both of which clearly have important implications for the environment. . . . Moreover, "nature" (water, air, minerals and vegetation in its uncultivated state) is regarded in traditional socialist theory as a free resource. This means there is virtually no accountability of the enterprise for the quantity of inputs used up in production. Because capital is cheap and readily available for . . . high priority industries, maintenance is poor and machinery gets run into the ground, which results in it working inefficiently, wasting more energy and raw materials and producing more pollution. . . . There is no incentive to introduce innovations which would result in less wasteful use of raw materials and energy, or reduce pollution.

Under this system, capital was rarely reinvested in production technologies or in improving efficiencies. Wars and past technical assistance from Russia, China and the US, created a network of factories that use technologies of drastically different vintages and qualities, many dating from the 1950s and 1960s. As a result, Vietnamese SOEs are now left employing very old, often obsolete manufacturing technologies. Pollutant emissions from industry in Vietnam are very high relative to state-of-the-art or even average international production technologies. Worker health and safety conditions are also quite severe throughout Vietnamese industry. A recent survey of working conditions showed that approximately 60 percent of workers were exposed to polluted air, excessive noise, and high temperature conditions in the workplace (Ministry of Health 1998). Although very little capital was reinvested in the maintenance or upgrading of production equipment in factories, even less seems to have been invested in environmental or worker protection measures.

Under Vietnam's economic reforms, subsidies to SOEs have been drastically reduced. Access to capital is now one of the most important issues facing SOE managers. Although certain SOEs still receive some special privileges on access to capital, competition for scarce funds has become intense. The cost of capital has risen rapidly (particularly compared to a previous situation of grants for factories). Capital is now often only available for investments that increase productive output and sales in the short-term. Factories receive loans if they can demonstrate that investments will have strong short-term returns. A tacit constraint on capital thus exists to exclude investments in pollution control equipment or longer-term investments in pollution prevention technologies.

SOE incentives and goals are also changing. Ministerial and external pressures are driving a transition from focusing on output, to focusing more on profitability. Competition based on price and quality—the new bottom-lines for industry—is creating pressure to lower costs and improve product quality standards. A new focus on export-oriented production is also driving demand for new production technologies. This modernization of Vietnamese SOEs is making them more efficient, and driving them to expand their production and resource use.

Locally managed state enterprises—those owned and operated by the People's Committee at provincial, city, district, and ward levels—are going through an even faster transition. The total number of local SOEs continue to fall as firms are closed or equitized. In 1987, there were 2,475 locally run SOEs. But by 1996, there were only 1,327 still in business. Here again, although total numbers of firms are decreasing, employment and output in this sector of the economy has shown growth of approximately 10 percent per year over the last several years, accounting for 17 percent of industrial output in 1997.

This transition, however, is hardly complete. Some SOEs still operate as if they are above the law and outside the market. For example, as one local government official complained, "because Lam Thao is state-owned, it feels it doesn't have to comply with government regulations." Tan Mai's decisions were explained by another local official who said, "because the factory belongs to Hanoi, the factory has a lot of power and they are not afraid of provincial authorities."

4.1.2 Multinationals and Joint Ventures

Vietnam has had a long and complicated relationship with foreign "investors." Russia and China provided "aid" during the American and French wars in the form of assistance in building factories. Lam Thao Superphosphates was built by the Russians, and Viet Tri Chemicals was built by the Chinese in the 1960s. U.S. aid agencies and multinational corporations similarly supported the industrialization of South Vietnam during the war. Tan Mai Paper was built by a U.S. firm in the 1960s.

Since 1990, the Vietnamese economy has come to depend increasingly on external investments in the form of Foreign Direct Investment (FDI) and official development assistance (ODA). Almost 40 percent of total investments in Vietnam now come from external sources (UNDP 1999). Until the Asian economic crisis began, Vietnam was one of the hottest investment locations in the world. Over $14 billion in FDI has been disbursed in Vietnam in the last ten years, with over 2,500 foreign-funded ventures approved by the government. Ho Chi Minh City alone has been pledged over $10 billion in FDI. FDI projects now account for 9 percent of Vietnam's GDP, and as of 1998, over $36 billion in capital had been pledged for current and future projects.

In 1987, the government passed the Foreign Investment Law that sought to facilitate investments in industrial development. Initially, most investments took the form of joint ventures. In the last few years, however, it has become increasingly common for projects to be 100 percent foreign owned. Output from foreign firms increased by over 20 percent per year between 1995 and 1998. From a tiny beginning in 1988, FDI has grown to account for almost 29 percent of industrial output.

It should be noted however, that FDI commitments decreased by 46 percent in 1998 due to the Asian economic crisis. In order to respond to this downturn, to improve the environment for investment, and to respond more generally to complaints from foreign investors, the government recently announced a range of measures (Decree No. 10 and Directive No. 11) to help speed the implementation of FDI projects. These include measures to simplify administrative procedures for foreign investors, lower land rents, and to institute a more even pricing system regarding salaries for employees of foreign invested businesses, prices for power, water and telecommunication services.

A new batch of MNCs has led the recent charge into Vietnam. As table 4.1 shows, corporations from countries such as Singapore, Taiwan, Japan, South Korea, and Hong Kong are the leading investors in Vietnam. Two of the case study firms were established by these regional MNCs. Dona Bochang was established by a Taiwanese firm, and Tae Kwang Vina was built by a Korean firm. Five of the six case study firms have thus had some support from foreign advisors or investors.

Table 4.1
Foreign investment in Vietnam

Country	Number of projects	Capital (Million dollars)
1. Singapore	236	6,740
2. Taiwan	618	4,970
3. Japan	301	3,800
4. South Korea	260	3,140
5. Hong Kong	201	2,680
6. France	108	1,800
7. Malaysia	58	1,369
8. United States	70	1,230
9. Thailand	78	1,109
10. British Virgin Islands	55	1,089

Sources: Business Times Singapore (2/1/01) and *Vietnam Investment Review* (2001).

Foreign investments have been concentrated overwhelmingly in manufacturing, with over 47 percent of capital dedicated to industrial projects. Hotels and tourism are the second largest sector representing 26 percent of investment capital. Industrial parks rank third with 13 percent of capital investments. Thus, fully 60 percent of foreign investments are focused on manufacturing or infrastructure to support manufacturing.

Almost all foreign investments in the early 1990s were in the form of joint ventures (JVs) with state-owned enterprises. The Vietnamese government was interested not only in attracting foreign investment, but also in fostering technology transfer, and in building the capacity of SOEs. Joint ventures usually involved the Vietnamese side providing the land, and the foreign side providing virtually everything else, resulting in a 70–30 split (70 percent foreign-owned and 30 percent Vietnamese). Foreign MNCs agreed to these arrangements as the price of entering a still highly controlled Vietnamese market. These joint ventures however, were often not very joint. As a director of Coca-Cola in Vietnam complained (personal interview—April 4, 1997), "We provide the money, the technology, and the management experience. The Vietnamese provide the land and the bullshit." Recently, Coke bought back the Vietnamese partner's 30 percent share of the venture to become 100 percent foreign-owned.

More and more MNCs are doing just this or establishing themselves at the outset as 100 percent foreign-owned. These are primarily manufacturing enterprises focused on production for export, and are located in industrial parks so they do not need to purchase land. Management disputes and corruption are also driving this trend toward 100 percent foreign ownership and management.

Most analysts assume that MNCs are more efficient than state-owned enterprises and Vietnamese private firms. With better technology and more advanced management practices, MNCs are also assumed to be less polluting. However, evidence from Vietnam and other countries in the region (see Aden, Kyu-Hong, and Rock 1999) indicates that this is not necessarily the case. Many of the early foreign investors in Vietnam imported secondhand equipment for their factories. This equipment, while new to Vietnam, was old and highly polluting, and often did not meet environmental standards.

Although more efficient managerially, MNCs may also be more exploitive than local firms. Some MNCs, for instance, seem to show little concern for the Vietnamese environment or the long-term sustainability of their operations. Taiwanese and Korean firms in particular have been the focus of criticisms in the Vietnamese press for their disregard for the adverse impacts of production.

4.2 Case Study Firms

The six case study firms I analyzed include centrally and locally managed state-enterprises, a joint venture, and a 100 percent foreign-owned firm. These firms vary significantly in not only their ownership structures but also their management systems, environmental staff, access to capital and technology, markets for their final products, relations to the state and local communities, and costs of compliance and noncompliance. And most importantly for this study, they vary in how they respond to regulatory and community pressures.

4.2.1 Dona Bochang Textiles

As mentioned, Dona Bochang Textiles company is a Taiwanese-run jointventure established in 1990 that exports over 80 percent of its products to other countries in Asia. Seventy-five percent of the capital for the venture comes from the Bochang Group, a Taiwanese conglomerate that operates four different businesses in Vietnam. The remaining 25 percent of the company's ownership is divided equally between the Dong Nai People's Committee and the Color Rubber Company (a provincial state-owned enterprise).

All of the equipment for the factory was imported from Taiwan, some of it new, but much of it secondhand. The Vietnamese side of the venture provided only the land and the government contacts for the factory. All of the raw materials for the plant are imported. The cotton comes from Pakistan and the chemicals come from Taiwan and Hong Kong. Eighty to 85 percent of its sales of towels—the company's main product—are for export. The factory employs 900 workers.

Dona Bochang's management is effectively run by the Taiwanese managers. The Director General of the factory—who is only at the factory about one day per week as he is also the manager of four other Bochang Group businesses—is nonetheless responsible for all major production decisions. For instance, to buy a new boiler or to install a new treatment technology, the staff must get approval from the Director General.

As of 1999, the factory had no real system for environmental management nor even a specific person assigned full-time to environmental issues. The Deputy Director is responsible for environmental problems along with many other responsibilities. The factory does however, now have an environmental committee made up of two engineers—the vice-foreman of production and the foreman of bleaching. Unfortunately, these staff have received little training on environmental or health issues.

Although the factory would not provide data on its profits, managers did reveal that they paid over $170,000 in taxes in 1997, indicating that the factory is making money. The market for its products remains strong despite the Asian economic downturn as they produce primarily low cost products. The factory also has good access to technology and capital through its Taiwanese parent company.

The factory's main rationale for not resolving its environmental problems is that the costs of retrofitting the plant to comply with air and water quality standards are now too high to make sense economically. Although the factory's costs of compliance are quite low for simple air pollution problems, it faces very high costs to comply with water pollution regulations or to install a new boiler. The cost of the wastewater treatment system alone is estimated to exceed $3 million.

So what has motivated the firm to take action? The managers of the factory asserted on several occasions that their main concern for their operations was their strained relationship with the local community. As one manager explained (Mr. Luat—3/13/97),

We are located in a residential zone so we are always thinking about complaints. We are not as polluting as factories in the industrial zone, but we have to be cleaner. The community complains more for a foreign joint venture. They want the government to be more strict on foreign companies.

We don't want to make any trouble with the people living nearby. We want a calm atmosphere for us to produce.

Dona Bochang is a company that can make environmental improvements if it is forced to. Community pressures—combined with DOSTE responses—now provide the motivations for Dona Bochang to act. However, this is also a company that imported secondhand, polluting equipment to Vietnam, built a textile printing and dyeing facility without a wastewater treatment system, and continues to lack internal procedures for environmental management. Community pressures must thus overcome a lack of internal processes or easy opportunities for technology upgrading. Thus far these pressures have been successful in motivating action only on relatively inexpensive technical solutions.

4.2.2 Lam Thao Superphosphates

Lam Thao is one of the largest state-owned enterprises in Vietnam. The factory was built with Russian assistance beginning in the late 1950s and began operation in 1962. Figure 4.1 shows villager housing with one section of Lam Thao in the background. The factory produces a range of chemicals including superphosphate fertilizer, NPK (nitrogen–phosphorus–potassium) fertilizer, and sulfuric acid for the Vietnamese market. Lam Thao controls approximately 50 percent of the domestic market for superphosphate fertilizers and a large share of the NPK fertilizer market, both of which are critical to the output of farmers across the country. The factory employs more than 3,000 workers.

Figure 4.1
Villager housing near the Lam Thao Superphosphates factory

Lam Thao is centrally managed. Formerly it was under the Ministry of Heavy Industry, but now is one of the flagship enterprises of VINACHEM, the state chemical corporation. Because Lam Thao is such a high-profile firm, much of the company's management comes from Hanoi. Lam Thao still receives preferential treatment from the government on a range of issues and receives some of the last remaining subsidies for transportation costs. Prices for apatite, pyrite, and coal—three of Lam Thao's largest input materials—are also set by the government below market rates.

Even with these subsidies, price controls, and preferences, Lam Thao appears to be only marginally profitable. With a near monopoly on fertilizer sales in the northern regions of Vietnam, the company has been able to charge sufficiently high prices to earn some profits. Lam Thao reported profits of just over 3 percent of revenue in 1997, and claims that it has been profitable for the last 14 years.

Lam Thao has a highly bureaucratized management structure with several layers of managers covering most issues. However, all major decisions seem to require approval from VINACHEM management in Hanoi. So even though the company has established an environmental team within its technical department, the midlevel manager responsible for handling environmental issues appears to have little power to implement environmental improvements on his own at the factory. As the vice-director of the factory explained (Mr. Loan—5/27/97), "We have to fulfill the plans assigned by the government, so if we spend too much time on environment we will have problems. But we do have a plan set by VINACHEM to be in compliance with the environmental law by 2005."

Lam Thao has had a difficult time trying to access new and cleaner technologies for its production processes. The company's real problem is its difficulty raising capital to buy equipment. Because of the massive scale of the factory, new production equipment can cost tens of millions of dollars. Getting access to state loans for large-scale investments has become increasingly difficult under market reforms. Lam Thao has also tried forming joint-venture agreements with foreign firms as a strategy to access capital and new technology. However, no foreign company has been willing to invest in Lam Thao.

Lam Thao is also faced with extremely high costs of environmental compliance. Simply upgrading the sulfuric acid line to bring it into compliance with air quality standards is likely to cost at least $8 million. The company has thus been trying to incrementally upgrade its equipment and to tie productivity gains to environmental improvements. As the vice-director explained (Mr. Loan—5/27/97), "Environment and production are closely related. If we pollute less, that means the efficiency of our production has improved. We have made the production changes because they raise the efficiency of the production process and improve the environmental situation for workers and the community nearby the factory."

Because of its strategic position, Lam Thao will continue to receive special privileges and is likely to survive the economic transition. But this privilege will also expose Lam Thao to more scrutiny from central government officials. Continued community complaints and demands for compensation will thus serve to motivate Lam Thao to reduce the impacts of its production on local villagers. However, environmental improvements are likely to come primarily during periods of technology upgrading when the company can devote scarce resources to solving pollution problems while also improving economic efficiencies.

4.2.3 Viet Tri Chemicals

Viet Tri Chemicals is also a state-owned enterprise managed by VINACHEM. The factory was built in the early 1960s with Chinese assistance and currently employs roughly 650 workers. Viet Tri Chemicals produces two main products—sodium hydroxide (NaOH) and liquid chlorine (Cl_2)—for the domestic market. The factory's main customers are its next-door neighbors such as the state-owned Viet Tri Paper Mill that is willing to pay above-market rates for Viet Tri Chemical's products because of long-standing relationships.

Viet Tri Chemical's management is similar to Lam Thao except that it is neither as prominent nor as profitable. It has thus been threatened with budget cuts and even closure since market reforms began being implemented in the late 1980s. Viet Tri does, however, still receive some special preferences from the central government such as reduced tax payments and subsidized water and other inputs. The government has also instituted a range of import tariffs to protect the company from Chinese and other foreign competitors.

Despite these supports, Viet Tri Chemicals has not made any profits for the last several years. In fact, as one manager claims (Mr. Tuyen—6/24/97),

As our output increases, our losses also increase. We expect more losses this year because input prices have gone up but we can't raise our prices.

Our price for sodium hydroxide (NaOH) is $330 per ton. Imported from China, you can buy sodium hydroxide for $220 per ton. In the United States it is only $100 per ton. Imported NaOH is taxed, but it is still cheaper than ours.

The Ministry of Industry and VINACHEM have requested assistance from the government give us some preferential electricity rates and to increase the tariffs on imported sodium hydroxide, but so far the government has not decided.

In response to these economic realities, the factory has sought to expand into other product lines. As mentioned, the plant manager has been talking lately about trying to produce cosmetics. The factory has also developed a laundry detergent and soap production line.

Access to technology and capital is a life or death issue for the factory. As one manager explained (Mr. Bieng—11/4/96), "Lack of capital is a chronic disease in Vietnam." Whether it tries to compete in sodium hydroxide and chlorine or to develop new products, it will need resources to invest in new production technologies. Upgrading the factory's sodium hydroxide production technology could cost up to $10 million. Figure 4.2 shows old polluting technology used in the factory.

With the very real threat of being shut down—due to the company's economic performance—the workers and managers of Viet Tri are nervous about anything that makes them look bad, such as pollution problems, negative impacts on the local community, or demands that the factory be moved. Viet Tri is thus quite sensitive to environmental complaints. The factory has established an environmental team with the Vice-Director of the factory in charge. This team, and this Vice-Director in particular, is one of the more effective environmental management groups I have come across in Vietnam. They understand the factory's environmental problems quite well, and have a strategy for resolving problems through equipment upgrades and efficiency improvements. The factory also has a program in place for rewarding employees who suggest environmental improvements that save the company money. In 1998, the company awarded workers 70 million dong (~$5,800) for suggestions that led to 800 million dong (~$67,000) in savings.

Viet Tri however, does not appear to feel much direct pressure from environmental agencies on pollution issues. Rather, it is concerned about its overall reputation and how this might impact future government loan decisions. The factory's director (Mr. Bieng—11/4/96) asserted that

We don't feel any pressure from the Ministry of Industry of VINACHEM on pollution issues. And no pressure from the National Environment Agency yet. We haven't received any written requests from NEA. The main pressure is from the community. The community sometimes complains directly to us. And sometimes the representatives of the community complain. Now people bring claims to the Center for Environmental Management or directly to the factory.

Viet Tri Chemicals is in many ways a tough firm to influence on environmental issues. It continues to cling to the idea of state supports, still hoping for state subsidies and protections. The factory also regularly plays the jobs card in the province, arguing that it must be supported to protect the 650 jobs it provides. And without major funding, the company is unlikely to be able to implement the pollution controls needed. But even

Figure 4.2
Old technology for chemical manufacturing at Viet Tri Chemicals

without a major injection of capital the company has been able to improve the efficiency of its production (thereby reducing wastes) and to improve its overall environmental management.

Viet Tri Chemicals thus appears to have both motivations and forced opportunities to respond to pollution problems. The pressure to reinvent the company and to drastically reduce costs has led to process innovations in production (in particular the change in electrolysis equipment) that also benefits the environment. Community pressures have been critical to these motivations.

4.2.4 Tan Mai Paper

The Tan Mai Paper Mill is a state-owned enterprise managed from Hanoi by the Vietnam Paper Corporation (VINAPIMEX). The factory was built in 1963 with assistance from the United States and taken over by the Ministry of Light Industry in 1975 after Vietnamese reunification. The company produces newsprint and other writing papers that it sells on the domestic market. Tan Mai employs just over 750 employees. Figure 4.3 shows Tan Mai rising up out of former rice fields.

Until a few years ago, Tan Mai was able to sell its newsprint to Vietnamese state-run newspapers at a fixed price. Poor quality and low brightness characteristics hardly mattered as these newspapers had no choice but to buy from Tan Mai or one of two other equally low-quality, state-owned paper mills. With the opening of the economy, however, newspapers were granted the freedom to purchase paper on the open market, including from foreign producers. The end to fixed prices and guaranteed markets was a painful one for Tan Mai. In the year of the transition, Tan Mai was unable to sell almost any of its paper. Foreign producers flooded the market with higher quality, whiter papers at much lower prices. Tan Mai is thus currently losing approximately 25 billion dong (about $2 million) per year. Because the company has no profits, it has not paid any taxes in the last several years. As of 1998, the company owed 12 billion dong (about $1 million) in back taxes.

If the Vietnamese government had cast its enterprises out onto the choppy waters of the "free market" to sink or swim, Tan Mai would likely have drowned in its own wastewater by now. But the government is not willing to let companies such as Tan Mai go bankrupt. For a number of

Figure 4.3
Industrial development (the Tan Mai paper mill) rising up from the rice fields

reasons, paper production has received priority from the government and it is currently working not only to protect Tan Mai and two other large pulp-and-paper mills in the country, but to actually expand these factories. The government thus has instituted import controls for paper products. For much of 1998 and 1999 it was illegal to import newsprint and other writing paper into Vietnam—a policy designed essentially to save Tan Mai from extinction.

Tan Mai is also still highly dependent on subsidies, price supports, and preferential treatment from the government. Tan Mai now buys pulp on the open market (imported largely from New Zealand), processes it into paper, and then sells the newsprint at a price set by the National Pricing Committee. Tan Mai does not pay for water used in its production process (although it uses millions of gallons of water), and the factory's waste treatment costs are subsidized. The factory's wastewater treatment plant itself was a gift from Sweden.

Tan Mai's worst pollution occurred during the period from 1994 to 1996 when the company was pulping Eucalyptus wood. However, in 1997, with the opening of the market, the company was forced to shut down the pulping line and switch to imported pulp, thus eliminating the most polluting parts of its production. Thus a market pressure—the fact that they could not sell the low-quality paper produced from domestic pulp—forced them to change their production, which turned out to be more environmentally sound. Tan Mai continues to use very old and inefficient production technology. The managers note that theirs is the only pulp mill in the world that uses a CTMP (chemico-thermo-mechanical pulping) system on bamboo wood.

Tan Mai claims to have a six-person environmental management team. This staff includes the operators of the wastewater treatment system and two process engineers charged with improving the efficiency of the pulping and paper-making machinery. Nonetheless, Tan Mai remains extremely inefficient in its paper-making. The managers report losses of approximately 7 percent of their raw fiber in their wastewater, compared to a world average of about one percent loss. The company also lacks a fiber recovery system so all of this lost fiber goes straight out the waste pipe to the waiting community outside. One estimate is that the community (which is surely much less efficient than a modern fiber recovery system) collects $70,000 worth of paper fiber per year. In the past, the factory has also had few incentives to reduce material use as water was free, raw materials were purchased at a set price, and energy was subsidized.

Management claims its biggest problem is access to capital to buy new equipment. They recognize that in order to compete with international paper producers, which Vietnam's membership in ASEAN will require them to do in several years, they will be forced to significantly upgrade their quality and lower their prices. But a state of the art pulp-and-paper mill can cost up to $1 billion. A pulp-and-paper mill built in northern Vietnam through a Swedish aid project cost over $400 million and is still not competitive. Although it would not make much sense for the Vietnamese government to invest this kind of money into a plant as old and inefficient as Tan Mai, it may happen. The government's commitment to developing a domestic pulp-and-paper industry has shown few reasonable limits.

Tan Mai seems to have a unique standing in the government. Because it has such strong supporters in Hanoi, there is little that the provincial or especially the local authorities can do to influence the factory. There have thus been almost no motivations for Tan Mai to improve its environmental performance. The company has been insulated from external pressures, but also denied the resources or opportunities to change. One hope for those concerned about pollution issues in the province is if the central government does decide to resuscitate and expand Tan Mai, it will be possible to include environmental improvements in new production technologies.

4.2.5 Ba Nhat Chemicals

Ba Nhat Chemicals was constructed in 1968 by the Hanoi city Department of Industry. The factory produces basic chemicals such as calcium carbonate powder for the local market. The factory uses 1960s Vietnamese technology (which is basically a local copy of 1950s Chinese or Russian technology) and appears not to have upgraded any of its equipment since the day the plant opened. The factory employs 200 workers.

Although extremely low-tech and inefficient, Ba Nhat claims to be profitable. The company sells most of its products to other small state-owned enterprises such as paint, rubber, and toothpaste manufacturers. The factory manager admits, however, that these profits are meager. And in fact he believes that if he had larger profits he would be able to ignore the complaints of local community members [although as we have seen this is certainly not always the case].

Ba Nhat is managed and run by the Department of Industry (DOI) in Hanoi. The DOI however, seems quite frustrated by the challenge of trying to keep firms like Ba Nhat open and at the same time trying to mitigate their negative effects. One agency official complained (Mr. Ty—7/5/97),

Most of the factories [like Ba Nhat] were built by Eastern Europeans so the technology is from the 1960s and 1970s. Most equipment is now obsolete. And there are no treatment systems in any factories. Now it is not worth it to invest in treatment systems for these enterprises. Old equipment is our most important barrier to solving environmental problems. The equipment that was installed before 1985 is obsolete.

Ba Nhat not only takes its lead, but also its protection from the DOI. So despite years of being pressured by the DOSTE in Hanoi to clean up its act, Ba Nhat was able to avoid either making costly changes to its production or being shut down. Ba Nhat has no environmental staff, and had no intention to improve its production while it was located in Hanoi. The factory's director explained the reasons he could not improve his factory's environmental performance (Mr. Hien—4/25/97),

> The factory is owned by the Hanoi Department of Industry. I can choose new technology if I have the money. But the government doesn't give any financial support to Ba Nhat. We have to use internal capital or go to a bank to get money for new investments.
>
> I do have to respond to the community complaints. But the most difficult issue is protecting my staff. I don't want to fire any employees. Our main incentive is the market. We only make changes that also benefit output and sales.

With such a small factory, and with such low profits, it is very difficult for firms like Ba Nhat to access commercial loans to buy new equipment. The factory manager explained simply that they could not generate enough profit to pay back even a low interest loan—which currently cost around 6 to 8 percent interest per year. Commercial loans are even more difficult to service as they often cost more than 1 percent per month (12–15 percent per year) in Vietnam.

The costs of Ba Nhat complying with environmental regulations is also high enough to make many decision makers think it would be better just to move the factory away from populated areas. Government officials believe there is little reason to invest in the plant as it exists now. The city and the factory manager realize there is no way to internally generate the profits to upgrade or improve the factory. But at the same time the city does not want to shut down the factory because of the lost jobs, tax revenue, and products that other city enterprises depend on. So, in a number of ways, Ba Nhat is one of the most difficult types of pollution problems to deal with in Vietnam—a classic example of a small-sized enterprise with few options or incentives to improve. Although Ba Nhat had motivations from a very well organized community, until its relocation was approved, the company had no access to resources or opportunities to upgrade production.

4.2.6 Tae Kwang Vina Shoes

Tae Kwang Vina is a Korean owned subcontractor for Nike Inc., the world's largest merchandiser of athletic shoes. Tae Kwang, which opened in 1995, is a subsidiary of a Korean conglomerate known as T2, which began producing Nike shoes in the 1980s in Korea, and as wages rose there, moved more and more of its operations to lower wage countries such as China and Vietnam. Tae Kwang's operations are highly labor intensive and the company now employs over 9,200 workers to produce approximately 6 million pairs of Nike shoes per year.

Tae Kwang specializes in producing Nike running shoes, some of which sell for over $150 per pair in the United States. One hundred percent of its products are exported (as it is illegal under its investment license to sell any of its shoes in Vietnam). The factory's top-quality shoes are exported to the United States, Europe, and Japan. Second-quality shoes are exported to other countries in Asia, the Middle East, and South America.

Tae Kwang is a highly profitable company. The company projected sales of $150 million and a profit of over $2 million in 1997 (only its second full year of operation). Costs of production are significantly lower in Vietnam than in Korea. For instance, the minimum wage in Korea is approximately $800 per month, while it is only $40 per month in Dong Nai province.

While Tae Kwang is run as an independent subsidiary of T2, its management is quite different than other private contractors. Tae Kwang's Korean managers are obviously responsible for day-to-day management of the factory, however, Nike oversees many critical aspects of production. Tae Kwang is in the top rung of Nike subcontractors and is considered a "strategic partner." These subcontractors get orders for the most technically sophisticated shoes that often also have the highest profit margin. Nike has quality control inspectors they call "manufacturing managers" in the plant every day analyzing production methods and quality performance. Nike sets the price it will pay for each shoe before it is made. And Nike sometimes even codevelops shoes with T2's in-house designers and engineers. Tae Kwang's relationship with Nike is even more complex regarding environmental and health and safety issues. As one manager explained (Mr. Im—3/14/97),

If Nike wants environmental improvements they recommend changes in chemicals or equipment. They suggest and advise us. Nike will give us a high-priced shoe order to motivate us to buy new pollution control equipment.

Nike sends an environmental inspector ten times per year to analyze air and water pollution, noise, chemical usage, how machines operate, etc. Nike came here to do an environmental audit. They studied air pollution, worker health, water pollution. They will send a report with requirements and dates when [improvements] must be completed. If we don't fulfill these, they will cancel contracts. If we do a good job, they will buy more high profit shoes.

Nike has many requirements for our company. Nike will lead our company and we will follow absolutely.

Nike's manager for labor practices in Vietnam explained the relationship in sometimes contradictory ways (Mr. Tien—4/12/97),

Nike has production and quality assurance staff in the factory every day. Factories have to sign a contract to follow the [Nike] code of conduct. If they want to manufacture Nike shoes they have to re-sign it every six months.

We can't tell them to change a solvent but we can talk about alternatives. We can't say 'don't use toluene.' But we can get them to work with the Nike chemical department. When technology is available, Nike pressures them to change technology. Then we discuss cost increases. We negotiate on investments for new machines. Nike can help them buy a new machine or let them charge Nike.

Nike can force companies to make changes. We can set deadlines. Nike can also pressure T2 to pressure Tae Kwang.

Another factory manager complained about the difficulty of working for the U.S. shoe giant (personal interview—10/7/97),

Nike kind of overreacts to [labor and environmental pressure] from the point of view of our managers. Human rights and environmental groups have attacked Nike. The media has blown things out of proportion. I don't want to criticize Nike, but they sometimes overreact to these things.

Sometimes Nike is too demanding. Some Nike staff don't understand footwear factory management. We can't make changes instantly.

But, we are selling an image. We want to keep the best image of Nike to get more orders. This is the real incentive for us.

Because Nike shares so much information with strategic partners such as Tae Kwang, the company has to take care of these relationships. As one Nike manager explained (Geoff Nichols—10/3/97), "It would be very difficult to sever a relationship with a shoe producer. Our strategic partners know too much. Tae Kwang has all of the latest Nike information. Our competitors would love to get that information." And at the end of the day, because Nike does not own and operate these factories, it cannot fully control these operations.

So for instance, although Nike had been putting significant pressure on firms such as Tae Kwang to improve their labor and environmental performance, even as of late 1998 Tae Kwang had no environmental staff or internal management team focused on environmental issues. Tae Kwang was essentially waiting for Nike to solve its environmental problems and then to transfer the technology and know-how. Tae Kwang felt this was a much more prudent strategy financially than trying to change technologies on its own, as Nike could help them with the engineering and management changes and then could help finance the implementation.

Tae Kwang also has a complicated relationship with the Vietnamese government. Tae Kwang hired the son of the Communist Party of Dong Nai province to work in the factory as a "problem solver." Tae Kwang contributes significant tax revenue to the government and helps employ thousands of workers. One government inspector explained (Mr. Thong— 8/6/98), "The Prime Minister issued a decree that we should facilitate production. So my main job, even though I am an inspector, is to give them guidance to improve. If after three inspections they have shown no improvement, we can fine them. But we don't want to shut them down because of the impacts on employment."

Tae Kwang is in a strong financial and technological position to change if Nike or the government requires it. As mentioned though, Tae Kwang is very close with the leadership of Dong Nai province and can thus avoid most state and worker pressures. The company obviously would prefer not to spend money on improving environmental and labor conditions if they do not have to. Nike is thus really the linchpin in influencing Tae Kwang to improve. But Nike itself first needs to be motivated to exert pressure on its subcontractors to change. This is where the role of external pressures such as NGOs and community members come in.

Citizen pressures on Nike in the U.S. and Europe have translated into motivations on firms such as Tae Kwang to improve. Nike is also helping to build the capacity of these firms to implement its code of conduct on environmental and health issues. Finally, Nike has through its contracting system essentially provided the resources for these firms to make technological changes to reduce pollution and workplace exposures.

4.3 Motivations for Action

Firms in Vietnam face a range of incentives to improve their environmental performance. The most direct of these are community pressures. However, firms can also be motivated by changes in market demands when international consumers demand products that meet ISO 14000 standards, "reputational" costs if they are associated with poor performance or negligence, and increased fines or compensation requirements for their pollution.

In a very direct sense, pollution can cost firms money. Solid, liquid, and gaseous wastes are essentially materials lost in the production process or materials converted into by-products that cannot be sold. Pollution thus represents raw materials purchased by the firm that are not incorporated into the final product. Pollution also costs money to manage (although not if a firm dumps their wastes illegally). If firms treat or control their wastes, pollution becomes a double cost. And if firms are required to pay fines or compensations, or are liable for long-term impacts of their emissions, then pollution can lead to a third set of costs.

Firms clearly do assess the costs and benefits of reducing pollution. Factory managers often mention the costs of capital (for making technology investments), increased operating costs (if new materials are required), the potential costs of disrupting operations, the potential impacts on sales, and the opportunity costs of using scarce resources for environmental investments. Costs of reducing pollution are then compared (whether formally or informally) to the costs of not reducing pollution. Costs of non-compliance include potential fines, compensations, and ultimately the threat of being shutdown.

Fines however, still do not present a major cost to firms. In Vietnam, as tables 4.2 and 4.3 show, fines are currently set at very low levels.

Tables 4.2 and 4.3 highlight that the highest fine in 1997 and 1998 in the provinces examined was approximately $370. Clearly these fines have

Table 4.2
Recent Fines in Phu Tho Province

Company	Year	Amount (VND)	Dollars
Lam Thao	1996	1,000,000	83.33
Viet Tri Beer and Sugar	1997	2,000,000	166.67
Viet Tri Chemicals	1997	2,000,000	166.67
Viet Tri Woodboards	1997	500,000	41.67
Viet Tri Paper	1997	2,000,000	166.67
Hong Ha Beer	1997	500,000	41.67
Viet Ha Material Company	1997	500,000	41.67
Phu Tho Tea Company	1997	500,000	41.67
Hoa Nam Beer	1997	500,000	41.67
Viet Anh Food Processing	1997	500,000	41.67
Huong Giang	1997	500,000	41.67
Pang Rim Dyeing	1998	5,000,000	416.67

Source: Centre for Environmental Management.
($1 = 12,000 dong in 1997–1998).

Table 4.3
Recent Fines in Dong Nai Province

Company	Year	Amount (VND)	Dollars
Dong Nai Breeding Company	1997	1,700,000	141.67
Transportation Enterprise	1997	2,800,000	233.33
Pig Breeding Enterprise	1997	800,000	66.67
TNHH Hung Hoang	1997	4,000,000	333.33
TNHH Hoang Loan	1997	500,000	41.67
Dong Loi Company	1997	4,400,000	366.67
TNHH Luc Sinh	1997	2,000,000	166.67
Ngoc Han Salt	1997	500,000	41.67
DonaFood	1997	3,000,000	250.00
Dona Newtower	1997	3,000,000	250.00
Thanh Binh Animal Food	1997	4,200,000	350.00
Luctie Company	1997	1,500,000	125.00
TNHH Viet Hoa	1997	2,000,000	166.67
DNTN Song Tien	1997	500,000	41.67
Thanh Cong	1997	500,000	41.67
Viet-French Pig Raising	1997	3,000,000	350.00
Interfood Company	1997	400,000	33.33
DNTN Nguyen Huu Tuan	1997	500,000	41.67

Source: Dong Nai DOSTE.
($1 = 12,000 dong in 1997–1998).

little economic impact on firms with multimillion dollar budgets. Every factory manager I interviewed acknowledged that the fines were too low to motivate changes. Others also noted that fines are not assessed to motivate compliance, such as if fines were charged per day, or in an escalating manner until a firm complied. Fines are levied one time per incident, and there appears to be little follow-up to monitor compliance. Fines are also not based on environmental impacts, but rather are based on visual determination of pollution impacts, and are set so low that all environmental violations are essentially treated equally. Fines appear simply to generate a small amount of revenue, rather than influencing firm actions.

Although fines are extremely low, compensations are sometimes set at levels that can influence firm decisions. Community demands for compensation have been increasing steadily since the passage of the Law on Environmental Protection. Communities now feel that they have legal rights not to be polluted, or at the minimum should be compensated for the impacts of pollution on crops and fish. Table 4.4 shows some of the recent compensations awarded to affected communities in Phu Tho province.

Lam Thao is the most striking example of compensation procedures. Lam Thao reported paying compensations of 636 million dong (about $54,000) in 1994, 540 million dong (about $46,000) in 1995, and 868 million dong (about $74,000) in 1996. So while the company paid less than $100 in pollution fines during this period, it was required to pay over $170,000 in compensations for environmental damages to local community members. Lam Thao managers themselves admit that they want to change production practices partly to avoid future compensations. Compensation demands are currently the strongest financial pressure on firms in some regions of Vietnam. The system of compensations however, is hardly ideal. Compensations are usually only paid after economic damages have occurred and can be proven by community members. Human health impacts do not usually warrant compensation. It is also interesting to note that compensations did not appear to be awarded in either Dong Nai or Hanoi, the locations of the other four cases.

Firms must also weigh the costs of waste management. Firms in Vietnam are currently not paying the full environmental costs of their pollution. Many of the environmental costs of production are being shifted onto local communities and the environment. As firms are not treating their wastes (as they should if they were in compliance with the Environmental Law) they are receiving an implicit subsidy from the government (paid by nature and local communities). Environmental management costs in the long-term should include staff salaries, and capital and operating costs for waste treatment systems.

It would thus seem logical for cost-conscious firms to seek ways to lower the current and potential costs of pollution—reducing wastes and increasing production efficiencies, and reducing long-term waste management costs. Rational firms should be continuously calculating and recalculating the costs and benefits of different courses of action. However, as we know, most firms do not take action to reduce pollution unless they are forced. This is partly due to the fact that evaluating the costs of pollution is given a very low priority in most firms. As long as wastes and waste management represent a small percentage of operating costs, most managers would rather ignore pollution issues. Until someone—regulators or community

Table 4.4
Recent Compensations in Phu Tho Province

Company	Date	Amount (VND)	Dollars	Reason
Bai Bang	Jan. 1995	10,500,000	875	Wastewater discharged into a fish pond and killed fish
Bai Bang	Sept. 1995	6,328,000	527	Wastewater pipe broke and damaged crops
Bai Bang	Jan. 1996	500,000,000	41,667	Support to Phong Chau district to compensate for environmental impacts of the factory.
Lam Thao	June 1995	98,000,000	8,167	Compensation to 8 communes in Tam Thanh district for crop damage from air pollution
Lam Thao	Aug. 1996	78,000,000	6,500	Air emissions during a breakdown destroyed rice fields in Cao Mai commune
Lam Thao	Sept. 1996	25,000,000	2,083	Wastewater spill into Cao Mai commune
Lam Thao	1997	260,000,000	21,667	Compensation for environmental damage when equipment broke down.
Viet Tri Chemicals	1996	42,000,000	3,500	Wastewater killed fish
Viet Tri Chemicals	1996	6,000,000	500	Chemical spill killed fish
Viet Tri Woodboards	1995	25,000,000	2,083	Wastewater killed fish
Vinh Phu Textiles	1996	14,424,922	1,202	Payment to the commune to repair a wastewater canal
Viet Tri Beer and Sugar	1997	10,625,000	885	Payment to repair a wastewater canal
Pang Rim Dying	1997	25,000,000	2,083	Wastewater discharged into a pond killed fish
Song Thao Ceramics	Sept. 1997	2,025,000	169	Air emissions damaged crops

Source: Centre for Environmental Management.
($1 = 12,000 dong).

members—forces the issue, firms are content to pay small fines or to shirk environmental responsibilities.

It is clear in Vietnam that the "market" alone is not motivation enough for firms to reduce pollution. Because the costs of polluting are so low, firms have few market signals on environmental performance. But even if the market were sending strong signals regarding environmental costs, which it is not, firms respond to much more than just market signals. So what then does motivate firms to take environmental actions that cost money?

Firms face a wide range of external pressures, including competitive pressures, regulatory mandates, and community pressures. Consumers, neighbors, workers, competitors, and a whole host of government officials (from the Ministry of Finance to the local Department of Science, Technology, and Environment) may attempt to influence the actions of a firm. However, environmental decisions get made within the confines of decisions over what to produce, how much to produce, and how and when to produce it. Profit motives, employment imperatives, and output targets, all normally supersede environmental decisions. Firms are also continually responding to the actions of their competitors, so production and sales strategies are often tied to other firms' actions.

Even for state enterprises, the bottom line is a bottom line, albeit a more fuzzy one. Ultimately, a simple costs and benefits framework appears to be employed by most factory managers when faced with environmental decisions. Even when analyzing compliance with government regulations, some factory managers calculate the costs of compliance against the costs of noncompliance (paying fines and compensations).

Decisions that affect pollution levels include choices about manufacturing technologies, management practices, resource allocations, and redesign of products and processes to reduce toxics use and emissions. Of course, investments in air pollution control equipment, wastewater treatment systems, solid waste management, and pollution prevention strategies are also critical to firm environmental performance. Decisions related to resource use efficiency (i.e., how water, energy, and raw materials are utilized) also influence pollution outcomes.

Sometimes production strategies clash directly with environmental concerns. For instance, some firms' production strategies—and why they came to Vietnam in the first place—involve emitting wastes untreated, thereby externalizing the costs of waste management. Lax enforcement of environmental regulations attracts some multinationals to locations such as Vietnam. Some state enterprises similarly assume they can get away with poor environmental performance as long as they appease their protectors in the state.

Factory managers report over and over that their number one priority is to survive financially. However, they also acknowledge the need to comply with local laws and regulations. Regulatory mandates, which are often quite weak in countries such as Vietnam, do serve as a pressure on firms to reduce pollution. As factory managers are embedded in social relations with government officials who regulate the actions of the firm, these managers often work to foster good relations with regulators.

Some firms are voluntarily working to prevent pollution through collaboration with the newly formed Vietnam Cleaner Production Centre (CPC). The CPC is a project of UNEP and UNIDO to provide free or low-cost technical assistance on pollution prevention and cleaner production strategies. These strategies focus on win–win methods for redesigning products and processes to be both economically and ecologically more efficient.

As we have seen, firms also respond to public pressures. Managers repeatedly express their desire to be good neighbors, or at least to avoid workers and neighbors complaining, protesting, or striking. As a manager at Viet Tri Chemicals claims, "Every factory wants to have a good reputation with the community around them." Firms are also sensitive to reports in the media and to other threats to the firm's reputation. A neighbor to Dona Bochang asserted, "Dona Bochang is afraid of publicity. The company does things to appease the peoples' complaints."

External pressures play a range of roles in influencing firm decisions. First, state agencies, communities, consumers, and NGOs can raise the financial costs of business-as-usual. Through a range of actions, these groups can influence firm financial calculations of costs and benefits of environmental actions. In the most effective cases, groups increase costs of noncompliance to the point where firms choose to take substantive action to reduce pollution. Compensations and fines are the most obvious examples.

A second role is to help focus the firm's attention on solving specific environmental issues. A manager at Viet Tri explained that "We set up [a system of monitoring stations] in 1990 because of complaints from surrounding community members." A third role is to shift the long-term perceptions and strategies of the firm. Nike managers explained for instance that NGO and community pressures regarding "sweatshop" conditions were affecting broader public perceptions of the Nike brand name and the image the company seeks to promote. A marketing study conducted in the United States found that teenagers associated the Nike brand with three things: (1) being cool, (2) being athletic, and (3) having poor labor practices. Realizing that environmental and labor concerns were undermining the long-term value of the company's marketing efforts, Nike took steps to improve conditions inside its factories, including the Tae Kwang Vina plant.

4.4 Environmental Decision Making

Research on western industries reports matter-of-factly that corporate environmental performance is influenced by "the 'stick' of government regulation, the 'carrot' of firms desiring a 'clean' public image," (Williams, Medhurst, and Drew 1993), and the long-term liability threats that both present (Flaherty and Rappaport 1991). Recent research by the World Bank on firms in developing countries points similarly to the importance of "reputational incentives," that is, pressure on firms to win and maintain a reputation as a "clean" or "green" company (Afsah, Blackman, and Ratunanda 1995). These pressures and market responses have supported a much heralded (but seldom proven) trend toward "green consumerism" and environmental marketing to increasingly aware consumers (Hayes 1991).

However, questions remain as to why some firms respond differently to similar government, market, and public pressures. Lawrence and Morell (1995) move beyond descriptive studies of "best practices" with the development of a model that seeks to explain firm environmental decision making. Their model argues that "green" firms are influenced by four key factors: Motivations, Opportunities, Resources, and Processes (MORP). Two conditions are necessary for their "MORP" model to work: management *Motivation* and proper internal *Processes*. Firms must be motivated, and then have processes in place so that actors inside the firm can respond to environmental problems. Two conditions are then supportive in the model: Opportunities and *Resources*. Ashford (2000) makes a similar argument in asserting that firms "must have *willingness, opportunity,* and *capacity* to change" (emphasis in the original). When firms have sufficient motivations, occasions for making changes, and proper internal decision-making processes in place, it is possible to solve environmental problems.

In an earlier study, Ashford and Heaton (1983) explored how external regulatory pressures influence internal firm decisions and can lead to technical solutions. They explain that the "need to comply with regulations often changes the nature of activity *inside* the firm: R&D may be redirected; new organizational units created; and a different set of alternatives may be faced. In addition, regulation changes patterns of activity in the world *outside* the firm by establishing new patterns of competition in a new market framework" (1983: 119). This analysis provides hints at how external pressures influence firm actions. The authors go on to assert that "a flexible and politically astute firm, able to influence and respond quickly to its external environment, can exploit regulatory opportunities," turning seemingly onerous regulations into competitive advantages (p. 127).

How firms respond to regulatory and other external pressures is obviously critical to understanding industrial environmental processes. As one OECD (1985) study has shown, a firm "may be immediately hostile to the regulations and spend time and money in circumventing them or disputing them in the courts; or it may adopt a constructive attitude and invest in research and innovation rather than lawyers" (OECD 1985: 67). In a country such as Vietnam, this "hostile" attitude can easily translate into firms bribing inspectors or using political leverage to avoid regulation. But here again, we are left wondering what motivates a firm to respond constructively rather than bribing its way out of regulations.

Gould, Schnaiberg, and Weinberg (1996) argue that companies only respond constructively under very specific conditions. First, there must be significant and sustained pressure on the firm. This occurs when a community realizes the company is not providing enough good jobs, tax revenues, or upward mobility for workers, and when the benefits of industrial activity are shown to be less than the environmental and health costs. Second, there must be equal enforcement of regulations in alternative production sites, thereby blocking a capital flight strategy. Third, modifications to production (demanded by the local communities) must be compatible with technological changes being considered as part of upgrading or modernization plans, which represent profitable changes or very small marginal costs to transform operations. Gould and his coauthors argue that these conditions are rarely achieved, even in advanced industrial countries.

Lawrence and Morell's MORP model, and Ashford and Heaton's analysis of external pressures on firm strategies are particularly relevant to this study's analysis of drivers of firm environmental decisions. As the case studies illustrate, motivations and capacities are key to firm actions, as are opportunities for firms to make changes.

Lawrence and Morell (1995) point out that creating opportunities for environmental change is critical to improving firm performance. When a factory upgrades its equipment, for example, this provides an opportunity to improve environmental practices, and creates a strong rationale for borrowing funds for environmental measures. Simply by upgrading production equipment most firms in Vietnam can capture environmental improvements. This is clear for Lam Thao and Viet Tri Chemicals. Almost any technology upgrade will result in pollution reductions (assuming output does not go up so much that it swamps the efficiency gains).

Although Viet Tri Chemicals has a very limited pollution control program, the company has made production changes that have reduced pollution. Switching from graphite to titanium electrodes in the electrolysis process helped the factory reduce its energy costs substantially, thereby lowering overall production costs. At the same time, this change reduced lead emissions and accidental releases of chlorine gas. Capital investments in equipment changes were justified by both environmental and economic benefits.

Over the last several years, largely due to community complaints, Lam Thao has also been forced to invest in pollution control equipment. The company built an acid-resistant wastewater canal, covered the canal with cement, then built neutralization tanks for wastewater, and installed air pollution control equipment. But more importantly, management of Lam Thao is now also looking at broader production changes for their environmental benefits. Incorporating environmental concerns into technology decisions provides critical opportunities for improvements.

4.5 Resources and Processes for Action

Once firms are motivated to take action, either by financial or nonfinancial pressures, the next critical issue is whether they have the capacities, i.e., the resources, technical knowledge, and institutional processes necessary to solve pollution problems. The most straightforward capacity issue relates to resources. Does the firm have the funds, or access to the funds, necessary to purchase new production or waste treatment equipment? And does the firm have the funds needed to pay and support environmental staff?

Second, if a firm has environmental staff, do these employees have the technical skills and the power in internal decision-making processes to promote pollution reduction efforts effectively? Can the professionalism and commitment of environmental staff overcome the imperatives of the accountants and managers focused on profits? Are environmental staff integrated into production decisions? Is the system for reporting environmental problems within the firm conducive to rapid decisions? Does the firm have processes in place and incentives for key actors such as engineers to find means to reduce pollution (financially or otherwise)? Are workers rewarded for finding pollution prevention options? Are environmental actions supported by top managers?

Barriers often exist that block firms from resolving pollution problems. These barriers can be informational, technical, organizational, or financial. Informational barriers, such as lack of data on the firm's use of raw materials and sources of pollution, and on alternative technologies can be overcome through appropriate research and development. At a minimum, firms need staff who can find and apply technologies to fit the firm's specific needs. Communication with outside information sources such as vendors and researchers supports this process. Firms also need staff who can accurately measure pollution problems and then make progress in solving them. Multinational firms often have more resources to perform these tasks.

Organizational barriers to pollution reduction occur when management structures marginalize environmental issues, when there is low worker involvement, and when engineers are provided limited incentives or rewards to care about pollution prevention. These barriers can be overcome by creating a professional environmental staff that are responsible for pollution reduction, and creating incentives for engineers and line workers to care about waste minimization. Top management commitment appears key to altering these organizational systems.

Accounting and financial barriers can also block pollution reduction efforts. For instance, firms often do not accurately account for material usage, and do not assign pollution control costs to the appropriate areas (overhead versus production). Many firms also fail to measure the full costs and benefits of production options, ignoring environmental costs and benefits.

None of the firms I studied had processes in place to respond strategically to environmental problems. Most firms still view the environment as a cost, to be dealt with by lower-level engineers. Even in the face of strong community pressures, firms remain reactive and defensive. However, if costs of pollution, inefficient resource use, and compensations continue to increase, some firms may learn that better environmental management strategies make business sense in the long-term.

4.6 Conclusions

Interviews with factory managers in all of the cases indicate that community pressures are a primary force motivating environmental actions in Vietnam. However, different firms respond differently to community and state pressures. State enterprises are insulated from community pressures and from state regulators in different ways than foreign multinationals. For instance, Tae Kwang Vina and Dona Bochang both complain that they receive more attention and complaints because they are foreign firms. However, they also have the resources (and different types of connections) to respond to outside pressures. Lam Thao, Viet Tri, Tan Mai, and Ba Nhat have close relations with state agencies, but they also have to continually justify that they are providing more benefits to society than costs. Each of these firms is vulnerable in different ways. Communities thus face different challenges to motivate regulation of these firms.

Community pressures can change the decision-making processes of factory managers. With increased community scrutiny, managers have less freedom to bribe an official or to cut a deal with a regulatory agency. Public and media attention increase the accountability of the firm, as well as the accountability of the local state agency. This helps block "fake" solutions and ensures government and firm promises are kept.

But even with these motivations, firms do not always respond constructively to these demands. Firms need the financial and technical resources to do something about environmental problems. State and community pressures are often not enough on their own. As Roodman (1999b) argues for firms in Vietnam, "on balance, it appears that public pressure can sway decision making at a company when money is not too tight." But without some access to capital and technology, it is difficult for even the most concerned managers to respond.

At the same time, it is critical to stress that the market alone is not enough to motivate environmental improvements in firms in a developing country. Firms in Vietnam do not operate as simple profit-maximizers in a theorized market-logic. They rarely make decisions based only on a calculation of the costs of polluting versus the costs of prevention or cleaning up. Other processes and pressures are central to their decisions. Market theories often ignore the underlying institutions that are not only key to the implementation of market-based instruments, but are at the heart of the processes that determine whether the state takes action to regulate firms, and whether firms respond positively or cynically. It is within existing institutional dynamics that firms make decisions about reducing pollution.

Firms are most likely to make environmental improvements when they feel sustained pressures (and particularly when these affect the firm's reputation), they have no easy escape option, they have the resources to invest, and when environmental improvements can be based around production changes that benefit the firm in other ways. Motivations, opportunities, resources, and processes are all key to environmental changes. In cases when these dynamics do not align, the challenges of promoting pollution reductions are even greater.

5

Motivating a Conflicted Environmental State

The party leads, the people control, and the state manages.
The state and the people implement together.
Communist Party slogans

When I first met Ms. Nhu in 1997, she seemed to represent the new face of Vietnam's government environmental programs. With a technical background and having completed a series of trainings on environmental monitoring and analysis, she was slated to become the head of the environmental analysis laboratory for Phu Tho province's Center for Environmental Management (CEM). Still in her late twenties, Ms. Nhu was poised to become a powerful young force for environmental management in her home province.

However, life inside an environmental agency in Vietnam is no easy duty. Staff like Ms. Nhu receive low pay (usually less than $50 per month) and face an uphill battle to do their jobs. The last time I saw Ms. Nhu, in May 1998, she was extremely frustrated by the state of CEM's new laboratory. More than a year after $100,000 worth of environmental sampling and testing equipment had been delivered to Phu Tho, it still sat unopened in boxes in Ms. Nhu's office. Although the equipment had been donated as part of a capacity building project funded by the United Nations Development Programme (UNDP), CEM simply had not been able to set up a room to adequately house the equipment. The cost of two air conditioners (needed to keep the equipment cool and dry) and a few sinks, was apparently too much for the province to contribute to support the startup of the lab. Ms. Nhu was thus instructed to read equipment manuals and attend more trainings until her boss could convince provincial authorities to invest a few thousand dollars in the lab.

As of late 1998, the province still had not made that investment. So when Ms. Nhu was given the opportunity to travel to the United States for an environmental training program, she participated in the training, then slipped away on the last day, never to return to Vietnam. A government official explained that she was now considered an "escapee" and would be prosecuted if she ever returned to Vietnam. The brightest young inspector from the Department of Science, Technology, and Environment (DOSTE) in Dong Nai province similarly disappeared while on this study tour, "escaping" an even more crucial environmental post in the south.

Much was certainly involved in these personal choices to abandon jobs and families. However, the decision by two excellent young staff to flee careers in environmental agencies certainly does not bode well for the future of environmental management in Vietnam. Ms. Nhu had basically given up hope of being part of an effective environmental agency. The unopened lab equipment, petty corruption, poor management from the senior staff (who knew little about technical issues), and low pay, had convinced her that CEM would neither provide a good career or likely be successful in protecting the environment of her hometown.

CEM of course is not unique. Throughout the Vietnamese government, as in many developing country governments, there is often both a reluctance and an incapacity to enforce environmental laws. Although the government has proclaimed a commitment to protecting Vietnam's ecosystems, workers, and urban environments, has passed a broad set of environmental laws and regulations, and has created national environment institutions and provincial enforcement agencies, the country continues to experience problems in the implementation of its environmental policies. A lack of funds, trained personnel, and political influence severely constrain the effectiveness of state environmental agencies. More importantly, contradictions and conflicts within the state—between developmentalist and environmental concerns—create incentives against enforcement of environmental regulations.

These conflicts can be quite pronounced in Vietnam. The Vietnamese state plays a significant role in both direct production (through state-owned enterprises [SOEs]), and the regulation of this production. The state has however, historically advanced a broader definition of development goals than simply GDP growth, promoting a rhetoric that includes protection of workers, development of rural areas, broad social equity, and protection of the environment. Delivering on these promises can raise conflicts and contradictions. Political stability and state legitimacy—which are continuously being renegotiated with social and political reforms—also remain issues of major concern to the government and Communist Party.

To be sure, CEM does not appear from the outside to resemble anything one would hope for in an environmental agency. The agency is rarely proactive, it is not particularly competent technically, and does not appear to carry much power or respect in the province. However, despite all of its weaknesses, it turns out that CEM does sometimes respond to public complaints about pollution, mediate conflicts, and negotiate regulation. And more surprisingly, these reactive functions can serve to pressure firms to reduce their pollution. Despite my generally poor impression of the leadership of CEM, they surprised me occasionally when they went beyond the required minimum, and actually worked to solve environmental problems.

As I found in my research, however, the staff of CEM and other regulatory agencies rarely act out of the goodness of their hearts or due to some deep-felt concern for the environment. Rather, the leadership of CEM was forced to take action on specific pollution problems due to an intersection of institutional responsibilities and public demands. And under specific conditions, despite all its weaknesses and conflicting interests, CEM was sometimes an effective regulator.

The opposite situation of course is much more common. Phu Tho province has environmental inspectors who virtually refuse to leave their offices, staff who are making money from kickbacks, and personnel who lack the basic technical capacities to conduct the inspections they are responsible for. Even committed staff face multiple pressures against enforcement from other state agencies, powerful state enterprise managers, and foreign investors.

This chapter seeks to examine why a government agency such as CEM with so many incentives not to enforce environmental laws, does actually regulate, and under what conditions this regulation is effective. I am particularly interested in what drives regulatory processes, and more broadly motivates and constrains state actions around the environment. In the contested terrain of regulation in a developmentalist state, it is critical to understand the processes and dynamics that can tip this balance towards enforcement.

Although this research confirms the constraints and contradictions of state environmental enforcement in Vietnam, it also shows that certain processes can motivate state actors to take measures to control pollution. In particular, in the six case studies, community pressures sometimes served to overcome state resistance to environmental enforcement, motivating local state responses to specific pollution incidents, pressuring environmental agencies to improve their monitoring and enforcement capabilities, and pitting different state agencies against one another. The current set of Vietnamese laws and institutions has fostered a system in which local environmental agencies are surprisingly responsive to community complaints, and are sometimes able to negotiate regulatory solutions between competing interests.

5.1 The Environmental State in Vietnam

Challenges and failures abound in environmental enforcement in developing countries. Vietnam begins its quest for development facing many of these common challenges. The country does not currently have cohesive environmental organizations, sufficient funds, or effectively trained staff to carry out many basic regulatory tasks. Environmental agencies are neither well insulated from capture, nor do they coordinate particularly well with other state agencies. Quite simply, few conditions are in place for effective enforcement.

The Vietnamese environmental state is however changing, and I would argue, strengthening its position and capacities. To put it in Barber's (1997) terms, Vietnamese environmental agencies are gradually strengthening their *human capital, fiscal strength,* and *legitimacy* (primarily through international aid programs and publicly visible environmental campaigns), but continue to struggle to strengthen their *reach, coherence,* and *autonomy.* Changes in laws, institutions, and public awareness have led to major changes in state *responsiveness* to community complaints about pollution. But the government has some distance to go to strengthen its environmental agencies.

5.1.1 Legal Basis of State Environmental Action

As noted in chapter 2, the Vietnamese government has issued a number of policy statements, laws, and decrees regarding the environment over the last 10 years. The Law on Environmental Protection (LEP), passed in 1993 (included in the Appendix), is an umbrella law that in many ways is more a statement of goals than a detailed plan for environmental regulation. Few features of the law are specific enough to be directly implemented. The LEP lays out general principles, and then describes state roles and responsibilities for advancing these goals. For instance, the law regulates the use of different natural resources (land, air, water, ecosystems), but does not specify the types of economic activities to be regulated or how they are to be regulated. The law states simply that "All acts causing environmental degradation, environmental pollution, or environmental incidents, are strictly prohibited" (Article 9).

The law goes on to assert that the government will set and control environmental standards, but does not specify these standards, or who is even responsible for establishing them. Article 50 of the law states that those causing damage to the environment or violating environmental regulations will be subject to administrative sanctions or face criminal prosecution. But here again, the law fails to establish how fines will be determined or imposed. Even on the issue of state responsibility, the law is quite vague. On the one hand, the law states that the Ministry of Science, Technology, and Environment (MOSTE) and the National Environment Agency (NEA) are responsible for implementing the law at the national level, and that People's Committees—and in particular the DOSTEs within the provincial People's Committees—are responsible for implementation at the local level. However, the law also encourages line ministries to establish their own environmental departments to manage environmental issues related to state-enterprises under their control, and notes that the Prime Minister can intervene in cases of "special severity."

An important aspect of the law is its statement on the rights of citizens to complain about environmental problems. Article 33 states that people who detect signs of pollution must immediately notify the local People's Committee. Article 43 states that the public has the right to complain or "denounce" state management of environmental problems. And Articles 49 and 52 establish that polluters must compensate those who suffer impacts from pollution. These statements of environmental rights, while still fairly vague, have served to legitimate public complaints and created a small window of opportunity for public participation in environmental issues.

Since the promulgation of the LEP, the National Environment Agency has drafted and the National Assembly has signed into law a series of decrees, directives, and circulars that flesh out the law, and create implementation instruments to realize the goals of environmental regulation and enforcement (listed in table 2.3 in chapter 2.) The most important are Decree 175 on the implementation of the Law on Environmental Protection and Decree 26 on fines for environmental violations. National environmental standards were issued in 1995, including standards on: ambient air quality; maximum allowable concentrations of hazardous substances in ambient air; inorganic and organic industrial emissions; pesticide residues; surface water quality; coastal water quality; groundwater quality; industrial wastewater discharge; and maximum permitted noise levels.

Five years after the birth of Vietnam's environmental laws and institutions, the Politburo of the Communist Party issued an interesting directive on environmental issues (No. 36-CT/TW, dated June 25, 1998). In this high-level policy statement, which reads almost like a Party self-denunciation, the Politburo frankly recognized that "the Law on Environmental Protection has not been strictly enforced," due to the fact that "the Party and the Government authorities at all levels are not fully aware of the importance of environmental protection," and "legal documents on environmental protection are lacking, overlapping and not consistent [and] investment in environmental protection remains low."

The directive asserts that in order to effectively implement environmental policies, the government needs both to change its policies and strengthen its enforcement. In particular, the directive points out the need to create greater awareness of environmental issues among the public at large by changing educational curriculum, giving people access to information on environmental conditions, and supporting public participation in environmental protection. The directive also proposes that the government more fully incorporate environmental considerations into development plans, and better promote pollution prevention strategies. The bottom line for the directive, however, is that the government must more strictly implement the Law on Environmental Protection and Environmental Impact Assessment law, and close down enterprises that cannot or do not reduce their pollution. This is a surprisingly open admission of the failures of current environmental laws, meant to show that the Party recognizes these problems, and that it plans to take action to improve environmental protection.

However, the reality of environmental management and enforcement is never as straightforward as laws and directives. Programs and policies that seem clear on paper can be interpreted and implemented in many ways. For instance, the Government has made environmental impact assessments (EIAs) a cornerstone of its environmental programs, but has not been able to implement them effectively. EIAs are written by consultants of varying knowledge and expertise and are usually reviewed by officials with even less knowledge or expertise. The review process is closed and has become a major source of corruption and conflicts of interests. There is essentially no quality control, or checks and balances on evaluation of the EIAs. And there are no procedures for public input or review of EIAs.

Even the consultants who write the EIAs (making around $2,000 to $4,000 per report) now often joke about how bad they can be. One story circulates in Vietnam about an institute that writes so many EIAs each year that they often just cut and paste chapters from one to another. This strategy came back to haunt them once when an EIA was submitted that discussed the food processing plant under review in the early chapters, but discussed a shoe factory in later chapters (pasted in from another EIA without changing the word "shoe" to "food"). Less than 10 percent of EIAs are ever rejected.

The government has also made much of Decree 26 and the fines it claims it will impose for violations of pollution laws. (These fines are discussed in detail in chapter 4.) However, actual implementation of this system of penalties varies widely in different locales, and the entire system of fines seems to have had little impact on pollution levels. One factory manager explained, "Fines only serve as a kind of warning not to pollute," while a local government official complained that, "Fines go to the state treasury. Losses are born by the local communities." Based on their current levels, fines do not even appear to be much of a "warning." The highest fine in the two years in the four provinces I studied was approximately $380 for a foreign firm and only $170 for a state enterprise.

Few people seem to believe government fines have any effect on the decisions of factory managers of medium and large plants. Clearly the most common fines of $50–$170 have little economic impact on firms with multimillion-dollar budgets. Many of the factory managers I interviewed acknowledged that the fines were too low to motivate changes, but at the same time complained that fines were not based on actual pollution levels (just visual determinations), and that the fines seemed designed to simply generate income for the government or for inspectors.

Although fines are extremely low as noted in chapter 4, compensations to impacted communities are sometimes set high enough that they can influence firm decisions. Community demands for compensation have been increasing steadily since the passage of the LEP. Communities now feel that they have legal rights not to be polluted, or at the minimum to be compensated for the economic impacts of pollution. The system of compensations, however, is hardly ideal. Compensations are usually only paid after economic damages have occurred, and only if the community members can prove that a specific factory caused the damage. Human health impacts do not currently merit compensation.

Actual enforcement is usually a matter of local politics and negotiation rather than a function of strict adherence to official policies. The Law on Environmental Protection grants local authorities the power of setting local regulations. Ho Chi Minh City, Hanoi, Dong Nai, and Viet Tri all have their own local standards, which they established while waiting for national standards to be promulgated. However, all of these regions except Ho Chi Minh City now largely follow national standards. Local governments have also experimented with other environmental strategies such as Ho Chi Minh City's creation of a "Black Book" of worst polluters in the city, Hanoi's attempts to relocate firms out of the city center, Dong Nai's promotion of industrial zones, and Phu Tho's efforts to negotiate regulations with large SOEs.

Both at the national and local levels, Vietnam has advanced essentially a traditional command-and-control regulatory system. The government has established standards, and is in the process of setting up the monitoring and enforcement systems to ensure compliance with those standards. This strategy depends on strong state enforcement capacity that can transmit instructions and incentives for compliance. Unfortunately, with the current system of EIAs, inspections, fines, and penalties, the LEP and accompanying decrees seem to create few incentives for compliance. This may simply be a symptom of new laws and institutions taking time to establish their capacity and authority. Or it may be that these policies were designed to respond to public demands without impeding economic growth.

Vietnam's long list of laws, decrees, and directives now comprises a fairly comprehensive legal framework for environmental protection. However, having laws on the books, and implementing those laws are two very different things. Issues of capacity, coordination, and conflicts of interest all rear their ugly heads when it comes time to move from policy to implementation.

5.1.2 Structures, Hierarchies, and Coordination

Ms. Nhu was, until her departure from the Center for Environmental Management, at the frontlines of the Vietnamese government's efforts to promote environmental protection. In each of the country's sixty-one provinces and its three largest cities (Hanoi, Ho Chi Minh City, and Haiphong), the government has established Departments of Science, Technology, and Environment (DOSTE), which are responsible for monitoring and enforcement of the nations' environmental standards and regulations. CEM is the environmental management division of the Phu Tho province DOSTE. And it is departments such as CEM that are largely responsible for influencing whether or not laws are implemented and whether or not the environment is protected.

The National Environment Agency (NEA), within the newly created Ministry of Natural Resources and Environment (MONRE), is officially responsible for environmental management in Vietnam, and in particular for the implementation of the Law on Environmental Protection. However, the NEA and DOSTEs are only two players in an extremely complicated patchwork of ministries, departments, and agencies operating at multiple levels who are responsible for regulating pollution. Table 5.1 lists some of the agencies involved in different aspects of environmental policy development and implementation.[1]

Table 5.1
Agencies involved in environmental policy

Environmental Activity	Government organization
Policy making	Communist Party of Vietnam Prime Minister National Assembly Provincial People's Councils National Environment Agency
Planning	Ministry of Planning and Investment Ministry of Finance Universities and Institutes Provincial Departments of Planning and Investment Line Ministry Planning Departments Ministry of Science, Technology, and Environment
Oversight	Ministry of Science, Technology, and Environment (MOSTE) National Environment Agency Provincial People's Committees Line Ministries
Implementation	National Environment Agency Provincial Departments of Science, Technology, and Environment (DOSTEs) Line Ministry Environment Departments

Source: Adapted from UNDP (1995).

In the six case-study firms, each was regulated directly by a provincial or city DOSTE. However, five of the six were also regulated at different times by the NEA or MOSTE. And four of the six also had to report to the Ministry of Industry on environmental issues. All of the firms were regulated by multiple ministries and departments, as table 5.2 indicates.

As the DOSTEs are the frontlines of environmental management (and the bottom of the hierarchy), I spent significant time inside each of the local DOSTEs interviewing staff about their regulatory work, and about incentives and barriers to enforcement of environmental laws. I also conducted interviews with NEA staff and environmental managers in the Ministries of Industry, Planning and Investment, and Construction.

Environmental management divisions were officially established in the DOSTEs in 1995, only two years after the creation of the NEA and the

Table 5.2
Regulatory and Promotion Agencies

Firm	Regulatory agencies	Promoting agencies
Ba Nhat Chemicals	Hanoi DOSTE Hanoi Department of Industry	Hanoi Department of Industry Hanoi People's Committee
Dona Bochang Textiles	Dong Nai DOSTE MOSTE (for EIA)	Dong Nai People's Committee Ministry of Planning and Investment
Lam Thao Superphosphates	CEM NEA Ministry of Industry	Ministry of Industry
Tan Mai Paper	Dong Nai DOSTE NEA Ministry of Industry	Ministry of Industry
Tae Kwang Vina Shoes	Dong Nai DOSTE NEA Dept. of Labor	Ministry of Planning and Investment Dong Nai People's Committee
Viet Tri Chemicals	CEM NEA Ministry of Industry	Ministry of Industry

passage of the Law on Environmental Protection. In 1995, there were only 120 government employees nationwide working within the DOSTEs on environmental issues. By 1998 there were more than 260 DOSTE environmental staff (UNDP 1999). Even with this growth, there are still only approximately four staff per province responsible for environmental management, which in most provinces is grossly insufficient.

During this same period, the NEA expanded from fifteen staff to over seventy employees and now has ten divisions. There is a proposal currently being advanced to expand NEA's staff to 350 employees, and to raise NEA to a general department within MONRE headed by a Vice-Minister. Proposals are also on the table to double or triple DOSTE environmental staff nationwide. This would begin to bring Vietnam in line with other environmental agencies in the region.

The growth of the Center for Environmental Management (CEM) in Phu Tho is a case in point. When I first visited their offices in the summer of 1994, CEM consisted of two rooms in a run-down building, containing little more than a few wooden chairs and a ceiling fan. The organization had no environmental monitoring or analysis equipment and little environmental experience or knowledge. However, by the summer of 1998, CEM was operating out of a new three-story building with more than ten offices, a conference room, and the still-uncompleted laboratory. Staff had been trained in using direct-reading monitoring equipment, and computers were operating throughout the offices. The DOSTE in Dong Nai experienced a similar expansion during these years, completing a new building of its own by late 1998, and expanding its environmental staff from three to ten employees.

The Law on Environmental Protection also requires line ministries to establish their own environmental management divisions. Most ministries have thus created nominal environmental divisions, some of which are much stronger than others. The Ministry of Construction for instance, has a much more developed environmental unit than the Ministry of Industry. The Ministry of Agriculture and Rural Development now has expansive environmental programs under its control (primarily relating to forestry and rural ecosystems). And the Ministry of Planning and Investment (MPI) now has a powerful environmental unit within the Department of Science, Education and Environment (DSEE). DSEE is the counterpart to NEA and MONRE on many development projects which have environmental impacts. Figure 5.1 shows a basic schematic of the government structure on environmental management, and figure 5.2 shows the key ministries involved in environmental issues.

This dispersal of environmental responsibilities to different ministries has both positive and negative implications. On the positive side, it has helped to "mainstream" environmental concerns into many aspects of government decision making, and has spread environmental responsibilities to officials closest to decisions that directly impact the environment. However, one major problem in this design is a resulting overlap of responsibilities and competition between ministries and agencies to control decision making over environmental planning, EIA review, and policy formulation. This dispersal of responsibilities without attendant mechanisms for resolving interministerial disputes remains a problem in many areas of environmental management.

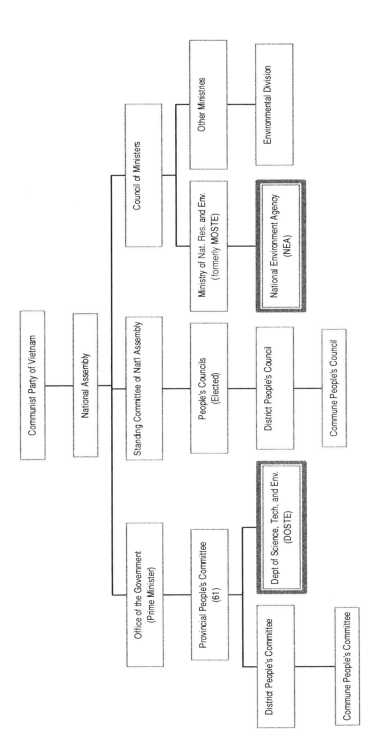

Figure 5.1
Government structure for environmental management

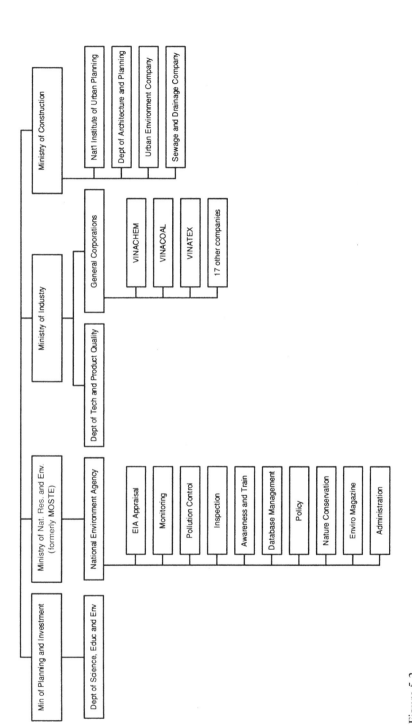

Figure 5.2
Ministries involved in industrial environmental issues

The organizational relations between the central government and the provinces around environmental management is also quite complex. The NEA was formerly a division of the Ministry of Science, Technology, and Environment and is now part of the Ministry of Natural Resources and Environment. Ostensibly the Departments of Science, Technology, and Environment were the provincial versions of the MOSTE. However, the DOSTEs do not report directly to, or take instructions from, the MOSTE. Rather, the DOSTEs are managed by the People's Committees of each province or city, which report directly to the Office of the Government and the Prime Minister. The NEA does attempt to guide the DOSTEs on implementation of national environmental policies, and occasionally hires them to carry out national inspection programs or EIA reviews.

As figures 5.1 and 5.2 should indicate, environmental institutions in Vietnam in general are organized along vertical hierarchical lines, and as such transmit information and orders up and down along these vertical chains of command. The different organizations and institutions responsible for environmental issues seldom pass information or coordinate among other institutions horizontally. So for instance, the provincial and city DOSTEs do not generally coordinate well with the National Environment Agency, and in turn the NEA does not coordinate well with the Ministry of Planning and Investment's Department of Science, Education, and Environment.

Even with the advent of advanced information technologies, environmental institutes and agencies seldom share information in Vietnam. Institutional tendencies towards vertical hierarchies and poor horizontal communication make coordination on environmental issues extremely difficult. As the World Bank has noted, environmental responsibilities are generally "unclear, suffering from the inherent danger of multiple responsibilities with inadequate accountability" (World Bank 1994: Annex 10). There also remain tensions between central and local environmental enforcement in Vietnam, and poor channels of communication and coordination between Hanoi and the provinces, and horizontally between provinces and cities (who presumably could learn much from each other).

In most cases, cooperation and information sharing remains poor between government agencies, institutes, and research centers. Information and data collected by a government agency can be requested through formal channels, with an official letter. However, this procedure sometimes requires months before receiving the data. Instead, researchers and agency staff report that they use personal connections to access information. The quality of personal relations often thus determines whether an organization can access information at all. According to some sources, environmental information sometimes does not flow well even between divisions of the NEA.

There are also limitations in the exchange of information from the local level up to the national level, even within a ministry. There are currently no clear mechanisms or protocols to ensure a reliable flow of information from the local to the national level. One of the reasons for the reluctance to share information is the perception that information, even if it has been generated under government contracts, should be controlled. People stated frequently that information is power, and that controlling information is one way for agencies and centers to earn additional income to support their staff and programs.

Numerous policy documents, consultants reports, and aid assessments discuss the weaknesses of current environmental institutional arrangements in Vietnam (see for instance, Thanh 1993, World Bank 1994, UNDP 1995, World Bank 1997, UNIDO 1998). Carew-Reid (1998) summarizes the institutional challenges of promoting environmental protection in Vietnam, including: limited separation of powers between the legislative, executive, and judiciary branches; limited coordination and integration between and within ministries, between the provinces and the center, and among the provinces; overlap in executing responsibilities; an overly centralized institutional structure; a complicated and inflexible institutional structure; limited monitoring of agency implementation; limited capacity of staff; limited resources devoted to environmental issues; failure to integrate environmental strategies into socioeconomic plans; and lack of follow-up on EIAs and environmental assessments. This combination of institutional and administrative problems has led to a situation where "procedures are cumbersome, administered by a relatively large number of staff at low levels of remuneration. Enforcement of laws and regulations is neither strict nor consistent, adding further to the uncertainties and unpredictability of administrative decision making" (Carew-Reid 1998: 14).

Decision making on environmental policy issues is controlled largely from the top of the political hierarchy. The Communist Party has retained

control of most aspects of policy development and controls key implementing positions at all levels of the government. This has created an
environment in which as Carew-Reid (1998: 13) argues "little creative authority has been devolved." The formal process of environmental policy
development in Vietnam generally involves the NEA drafting a new policy
(with the assistance of university researchers or other government institute
staff), then sending it up the chain to the Ministry of Science, Technology,
and Environment. MOSTE reviews the policy and then submits it to the
Office of the Government. When the Prime Minister has signed off on the
policy it can be submitted to the National Assembly for approval, or issued
directly from the Office of the Government.

However, even within this formal process, the Communist Party continues to exert control over most key decision points, including the National
Assembly, executive offices, and the ministries and departments that review a policy proposal. The National Assembly, which was until recently
a rubber-stamping body, still only meets twice per year for short sessions,
and rarely votes against official party decisions. The National Assembly
would score very low on a legislative professionalism index (Agthe, Billings,
Marchand 1996).

This structure of course has costs. As one critic (Porter 1993: 64) argues,
"the price of a centralized political structure under the domination of the
party has been an overburdened leadership group, rampant abuse of power,
popular alienation, ineffective policy-making on many issues, and an inability to respond administratively to popular needs." Dryzek makes the
point more generally for centralized environmental agencies, arguing, "hierarchical systems necessarily obstruct the free transmission of information
that is essential to the effective solution of nonroutine problems. In the context of ecological problems no less than elsewhere, verified truth relevant to
policy does not exist. Theories in the hands of the apex of administrative
structure must always err to a greater or lesser degree. But hierarchy re-
sists the institutionalization of trial and error that this recognition neces-
sitates. . . . Hierarchy may be adequate for the coordination of routine
tasks, but not for complex and variable problems" (Dryzek 1994: 182).

In a major, multiyear study of environmental issues in Vietnam, the
World Bank (1994) made one of its highest priority recommendations
the coordination of environmental policy and implementation through

"the establishment of a supra-ministerial body or council to be chaired by at least a Deputy Prime Minister with MOSTE serving as the secretariat." Three years later, in another report, the World Bank (1997: vii) once again put the establishment of a cross-ministerial consultative committee to coordinate between MOSTE, the Ministry of Planning and Investment, the Ministry of Finance, and the Ministry of Industry as its top priority recommendation. However, to date, no such coordinating committee has been established, and lines of decision making around pollution issues remain unclear. Laws are now on the books and institutions are in place. However, for virtually every major pollution issue the lines of decision making and enforcement responsibility become hazy when it comes to specific responsibilities and enforcement.

5.1.3 State Capacity

The actual implementation of environmental policies continues to be stymied by weaknesses in agency capacities to monitor and enforce laws and regulations. Studies in developing countries in Asia, Africa, Eastern Europe, and the Americas repeatedly discuss problems of inadequate staffing and funding, lack of scientific knowledge, professionalism, institutional capacity, coordination and cooperation, limited independence, and the overcentralization of environmental responsibilities in weak national environmental bureaucracies (see for instance Dwivedi and Vajpeyi 1995, Eder 1996, Desai 1998, Sapru 1998). The DOSTEs that I studied were all understaffed, underfunded, and undertrained to carry out the long list of duties assigned them. Not surprisingly, the heads of the DOSTEs complained regularly about these weaknesses.

A number of capacity building projects have been sponsored by international agencies and donors over the last five years in Vietnam including programs to build DOSTE capacities in Hanoi (funded by the Vietnam–Canada Environment Program [VCEP]), Dong Nai and Phu Tho (funded by the United Nations Development Programme and implemented by the United Nations Industrial Development Program), Ho Chi Minh City (funded by the Singaporean government and UNIDO), and the National Environment Agency (funded by the Swedish International Development Authority and VCEP). The World Bank has also funded a four-phase project on building Vietnamese capacity for industrial pollution prevention.

Capacity of course is more than just technical skills, staff numbers, or access to new equipment. The unopened lab equipment in Phu Tho is only the most glaring example of the difficulties of building capacity. State institutions that are weak or lack legitimacy face severe constraints on their ability to intervene forcefully in socioeconomic activities (such as the regulation of industry). State agencies are often only as powerful as the degree to which they can distribute favors or extract resources. As one UNDP program officer explained (Ms. Ly—12/11/98), "while there are now many people in the Vietnamese government who can talk about environmental issues, there are no effective champions for moving environmental programs forward. The current leaders cannot organize, mobilize, and lead a movement in the government. . . . Vietnamese experts do not have good coordinating capacities. We need a longer-term strategy for developing a younger generation of environmental leaders." Other characteristics are also critical to environmental enforcement such as state coherence (the degree to which agencies and actors in the state agree to work toward shared objectives), autonomy (the extent to which the state can act independently of external forces), legitimacy (the strength of the state's moral authority), reach (the state's ability to get things done in society), and responsiveness (the extent to which polices meet the needs and grievances of citizens) (Barber 1997).

Coordination and decision making between agencies and departments can also be critical to the implementation of environmental policies. Decision-making authority can be both too highly centralized and too fragmented (Desai 1998). Centralization can lead to problems when regulators do not fully understand problems on the ground, trade-offs that need to be taken at the local level, or the power of local vested interests. Rigid hierarchies are also less capable of learning from local experiences and experiments in implementation. On the other hand, completely localized policy implementation can lead to problems of corruption, poor coordination between locales, and poor information sharing. With the globalization and intensification of environmental problems, and the increasing speed of changes in industrial practices, state administrators are hard pressed to keep up with industrial and environmental changes. Top-down technocratic decision-making processes are particularly limited in their capacity to respond to new economic dynamics and environmental problems (Giddens 1990, Beck 1992).

Even on the most straightforward capacity issues—those related to funding for agencies—the Vietnamese government faces challenges. The government has set a target of spending 1 percent of GDP on environmental investments by 2010. However, the budget for environmental programs is currently only approximately 0.03 percent of GDP, a fair distance from the stated goal. Large amounts of foreign aid money have been flowing into Vietnam for capacity building purposes. However, this has led the Vietnamese government to choose not to invest much of its own money in agency staff training or human resource development.

5.1.4 Conflicts in the State

As numerous analysts have argued, the state is often an arena in which competing interests attempt to advance their goals and objectives. Gould et al. (1996) argue that the state continually faces conflicting interests and goals around environmental issues. At the most basic level, the state must both facilitate economic development and protect the social and economic structures that support society. This means both promoting industrial expansion and trying to protect "livelihood" and "quality of life" issues including clean air and water. Sometimes this takes on a "zero-sum" competition for access to natural resources (such as use of forests for timber companies or national parks). Other times the state simply is not capable of effectively enforcing environmental regulations in the face of powerful corporate interests.

A continuing theme in studies of environmental enforcement in developing countries is the conflict between economic and environmental goals. The globally hegemonic economic development strategy of export-oriented industrialization is often fueled by the exploitation of natural resources (which leads to degraded environments) and the expansion of industrial manufacturing (which leads to urban pollution). One recent survey of these issues (Desai 1998) showed that, in the ten countries studied, economic growth took precedence over environmental protection, and economic policies and programs led to serious impacts on the environment. In every country analyzed, environmental ministries and agencies were much weaker than economic and planning agencies.

In environmental issues it becomes quite clear that "the state" is not a monolithic entity. Conflicts exist within state agencies regarding the costs and benefits of environmental regulation, and between state agencies regarding the regulation of activities with environmental impacts. State agencies face a range of costs and benefits related to environmental enforcement, including the financial costs of building agency capacity and carrying out inspections, as well as the political costs of upsetting factory managers, other government officials, and local constituents. There are major political and economic costs associated with closing down a polluting factory (which is why it is so rarely done). On the other hand, there are also political costs to ignoring community complaints and allowing local environments to be degraded. In the minds of government officials, there are potential costs to both overregulation that scare off foreign investors or harm the economy, and underregulation that allow destructive activities that undermine the local resource base and quality of life. Ultimately, it is a combination of costs that must be compared, ranging from the costs of environmental protection programs, to the economic impacts of these programs, to the effects on tax revenues, all compared with the benefits gained from environmental and health protections.

The World Bank (1997), for instance, has estimated the current economic costs of health impacts due to industrial pollution in Vietnam to be approximately 0.3 percent of GDP and rising fast. The Bank further estimates that health costs of pollution could reach hundreds of millions of dollars per year, or 1.2 percent of GDP by 2010. In this light, current environmental spending is an order of magnitude lower than the social and ecological costs of pollution.

There are also direct personal costs and benefits of enforcement within state agencies. Low salaries make it not only tempting, but sometimes a matter of survival to earn extra money from regulated parties. Getting a reputation for being a strict inspector may also raise unwanted attention and pressure from higher officials. Rational agency staff are thus hypothesized to minimize these costs by regulating firms that either can afford to pay or who do not have strong local support (Peltzman 1976).

Corruption is currently a serious problem throughout the Vietnamese government. Although there is no open documentation on this subject, it is widely held that inspectors take bribes to supplement their salaries. Transparency International ranks Vietnam in the top 10 percent of countries in perceived corruption. This can range from small "gifts" such as free dinners to major cash exchanges. The most common form of corruption seems to involve kickbacks for government approvals and contracts. For instance, it is common to hear stories about how managers on international aid projects can build themselves a new house by skimming a percentage off project budgets. One Asian Development Bank (ADB) staff person in Hanoi asserted (personal interview—12/23/98) that up to 30 percent of project budgets can disappear in kickbacks for employees and contractors. Even the former Prime Minister complained that when the government tries to give 100 dong to the people, 80 dong gets appropriated by corrupt officials on the way down.

There is also an issue of the autonomy of state agencies and actors. A state environmental agency cannot be "too close" to a firm, or it will not be able to independently monitor and enforce laws. By this, I mean the environmental agency should not be either fully captured or overly influenced by either a firm manager or other state officials with interests in the promotion or protection of the firm (such as the Ministry of Industry or the Ministry of Planning and Investment). These conflicts are particularly stark in regards to state-owned enterprises. When the state is both the polluter and the regulator it is difficult to overcome agency conflicts and advance normal regulation. The issue arises for example when the NEA has to deal with a pollution problem at a Ministry of Industry-affiliated factory. Although MOSTE is on the same administrative level as the other line ministries, clearly NEA is not powerful enough to force the Ministry of Industry, or even the environment department within MOI to do something it opposes. And as mentioned, there is no established system within the Vietnamese government for arbitrating interministerial disputes around environmental issues.

The example of trying to shut down factories is particularly telling. In 1995, NEA released a "blacklist" of thirteen firms nationwide it felt should be shut down because of the severity of their pollution. These firms were very clear polluters, with little hope of ever installing the technology to bring the firms into compliance. The reaction from other ministries and agencies however, was so strong that NEA's plans were blocked. To date, not one of the factories on NEAs blacklist has been closed.

To put it bluntly, the agencies charged with industrial development and environmental management in Vietnam—as in most countries—just do not get along. This is not simply an issue of poor coordination. This goes to the level of rivalries, competition for resources and power, and direct conflicts over goals and objectives. Ministries and provincial departments have ongoing feuds and power struggles, even among and within their environmental divisions. For instance, in each of the DOSTEs, the relationship between the inspection division and the environmental management division was less than ideal. And relations between the DOSTEs and NEA were even worse. As one UNDP official asserted, the NEA even "sometimes works to undermine local level initiatives such as the Environment Committee in Ho Chi Minh City" (Johnson—12/10/98). And relations between the NEA and other departments and agencies in Hanoi are also rife with political problems and competition.

5.2 Inside the Frontline Environmental Agencies

The frontline environmental agencies—the DOSTEs and CEM—face a continuous struggle to enforce Vietnam's environmental laws and decrees. Agency staff who are committed to enforcing the law face internal and external pressures against enforcement. Weaknesses in capacities and motivations of other staff further impede enforcement. However, under certain conditions these barriers can be surmounted, and regulation can occur. A series of interviews with agency staff in Phu Tho, Dong Nai, and Hanoi, the three provinces of the case study firms, provide some insights into these challenges and successes.

5.2.1 Phu Tho DOSTE and CEM

Phu Tho province has undergone a number of changes in environmental management over the last few years. The province was actually created in 1996 when a larger province—Vinh Phu—was split in two. At that time, the responsibilities for environmental management were also split between the two provinces, and staff were spread even more thinly across the new agencies. The Center for Environmental Management, with a few key staff missing, became the environmental agency responsible for Phu Tho. A reorganization a year later made the DOSTE, which is above the CEM, responsible for all environmental inspections and fines, with the CEM responsible specifically for environmental monitoring.

The Inspection Division within DOSTE now carries out inspections and responds to citizen complaints. Mr. Khanh, the director of DOSTE, explained the department's responsibilities and activities (personal interview—6/25/97).

There are three kinds of inspections: regular inspections which should be carried out every quarter or every six months; sudden inspections, where we give no warning to the factory; and, complaint-based inspections. Companies are informed five to seven days before the inspection. The sufferer of the pollution is also informed. In the inspections we try to inspect the real situation of the pollution. However, in the past we only used an 'intuitive' method for analyzing the pollution, we couldn't do any sampling. Now though we take samples and send them to Hanoi for analysis. The factory has to pay for the sampling costs.

Based on an analysis of the sampling results, compared to the standards, and using Decree 26 on fines, we decide whether to fine the firm or certify compliance. Fines go to the Phu Tho treasury. The factory can appeal the fine if they don't agree with it. But the factory still has to pay the fine first and then appeal afterward.

Our other responsibility is when the actions of a company affect a community. We work with the company to set the compensation for the community. We basically arbitrate between the two sides.

The province cannot shut down a centrally run state-owned enterprise. Only the Ministry of Industry or the Prime Minister can close a large factory.

The most common environmental inspections follow community complaints. Until the DOSTE assumed its responsibilities, the Center for Environmental Management was the lead environmental agency in Phu Tho. Mr. Hung, the director of CEM explained how the agency handled complaints, inspections and compensations in 1995 and 1996 (personal interview—11/5/96).

In the last two years there have been several cases of pollution and compensation. In one case there was a wastewater release [from Viet Tri Chemicals] that killed fish in a fish pond. The factory first claimed that the death of the fish wasn't their fault. CEM measured the pH of the water and was able to prove that the factory's wastewater had killed the fish.

Lam Thao has had two recent cases of compensation. In one, the factory emitted hazardous gases which burned 30 hectares of rice crops up to eight kilometers from the factory. After the incident, the local people called CEM and asked us to come out to verify the impact. We helped negotiate the compensation—which was 78 million dong for one commune and 20 million dong for another commune. The value of the lost crops were probably higher than that, but that was the compensation the company was willing to pay.

I have to explain to the communities that they can't get the full compensation because the firms use old technology and need to keep operating for employment reasons. Sometimes communities can't prove that environmental damage is caused by a specific factory. In those cases, they don't get any compensation.

Companies are worried about community members gathering in front of their factories. They don't like protests in front of the factories. These just began recently. We also are worried that these gatherings will spread if other factories don't clean up their pollution.

The public has more environmental awareness and wants the Environmental Protection law to be enforced. The public realizes that there is damage and costs from pollution.

In a follow-up interview six months later, Mr. Hung explained (personal interview—6/27/97) that although CEM is now responsible for environmental monitoring, because they do not have a large budget or a proper laboratory, they only do monitoring when community members complain, and compensations only occur through negotiations between the state, firm, and community members.

We choose which factories to monitor based on public opinions and environmental problems. Factories with bad reputations get monitored. If fish are killed, or a well is contaminated, we take samples. We can't do regular environmental monitoring. We don't have the staff or the equipment. So we only monitor when there is a complaint or a problem.

The amount of compensation is set by negotiation between the factory and the community. If we force the companies to pay compensation in accordance with the value of the community loss, the factories will have a very hard time. They will lose money.

For Lam Thao, we studied the pollution for three months in eight different communes. If we had suggested the compensation amount based on past losses, Lam Thao couldn't have paid. Lam Thao agreed to pay some compensation, but paid it to the District People's Committee, not to the communes. The community actually got very little money and complained to us again. After that we decided that all future payments must go directly from the polluter to the sufferer.

Thus, while Phu Tho has official inspection procedures, currently almost all monitoring and enforcement is driven by community complaints. Fines remain so low that they have little impact on pollution levels, but compensation demands from polluted communities can result in substantial costs to firms. All fines and compensations are negotiated between the state, firm, and community.

5.2.2 Dong Nai DOSTE

Industry in Dong Nai province has been expanding at exponential rates, growing by over 30 percent per year in the 1990s. There are now over 330 joint venture and foreign-owned enterprises in the province (with a total investment capital of $5.25 billion). At the same time, pollution levels have also been growing disturbingly fast. According to an internal government report, the Dong Nai DOSTE received roughly sixty complaints about pollution problems per year between 1994 and 1996. However, in 1997, the DOSTE inspection division reported receiving over 100 complaint letters in the first six months alone. These complaints came from community members living near factories, workers inside the factories, and even neighboring factory managers.

Community members send complaint letters to many different government offices in Dong Nai. However, at some point most complaints end up on the desks at the DOSTE, as they are ultimately responsible for pollution problems in the province. The DOSTE's current inspection system is similar to Phu Tho in that most inspections to date have been motivated by community complaints.

The Dong Nai DOSTE themselves complain that they lack the equipment and staff to monitor factory emissions, and more importantly, that they do not have the power necessary to influence firms to reduce their pollution. One DOSTE inspector explained that "in terms of the law, every factory is equal, but in reality older factories and state enterprises get treated differently. We give state enterprises more time to comply with the laws. New factories are easier to manage. For state factories with difficult emissions, we usually make a report and send it to the People's Committee, asking for suggestions for a solution. Decisions on these factories are made from the top. But each factory is different" (Mr. Tien—4/9/97).

The process of inspections in Dong Nai was explained as follows (Mr. Lang—6/6/97):

When a complaint letter comes in we establish an inspection group. It usually has seven or eight members, including people from DOSTE, the ward People's Committee, the district or province, and sometimes even a reporter. Sometimes though, we don't assemble inspection teams, just members of DOSTE go out to investigate.

The principle for solving disputes is negotiation. The factory and the people should agree together on the settlement. If they can't agree then we have to take some action like setting a fine.

The management of environmental problems is not particularly smooth within the DOSTEs. Although more closely connected than the CEM and

DOSTE in Phu Tho, the different divisions in the Dong Nai DOSTE also have conflicts and limitations, and most regulation appears to involve a process of negotiation first within the government agency itself, and then between the government and the firm. As Mr. Lang explained (6/6/97),

The integration between the Inspection Division and the Environmental Management Division within DOSTE is not good. There is poor coordination and the two divisions don't seem to understand each other. The Environmental Management Division only gives the test results. The Inspection Division makes the decisions on action.

There are also problems of authority in solving pollution problems. Some districts give responsibility to the health division, some to the economic division. There are no rules for the responsibility. The system should divide responsibility appropriately. Now it is just based on investment capital (big projects are handled by MOSTE, small projects by the DOSTE). This creates problems about administrative borders and for pollution between province.

Under Decree 26, the DOSTE can fine factories. However, the fines are too low, and companies are not afraid of the fines. Factories are only afraid of publicity through the media. They are worried that sales might go down. When you apply Decree 26, you have to be very flexible. You cannot strictly apply the fines. You have to consider social effects.

The DOSTE has the right to temporarily shut down factories for up to five days. For a longer period, we have to go to a higher level. So far, we have never shut down a factory.

There are sometimes conflicts between inspectors and foreign managers. If the inspectors are very strict about environmental enforcement, then it will have a negative impact on the economy. Foreigners come here to produce, and are not happy if they have to pay fines. Especially the Taiwanese and Koreans. We are afraid if we are too strict we won't get other investors from Taiwan and Korea. If you want the investment capital, you have to sacrifice some things.

Against these odds, the head of the inspection division reiterated how important public complaints are to the inspection process (Mr. Tien— 6/9/97), "If the complaints of a community are very strong, that factory will be inspected first. We have too many factories to inspect, so we prioritize based on complaints. The compensation process starts from complaints. If people lose crops but don't complain, then there is no compensation."

The director of the environmental management division explained further how delicate regulation can be (Mr. Het—8/1/98),

Every normal inspection in Dong Nai is preceded by a complaint. The DOSTE did conduct a series of inspections for NEA in 1997, but all of our other inspections are based on the complaints of people and their letters. Normally, the DOSTE conducts around 200 inspections per year. Fines are set by the inspection division based on Decree 26. The DOSTE can keep up to 30 percent of the fine if it submits detailed plans for using the revenue. But even this much isn't enough to cover the costs of sample analysis. The level of the fine isn't very important. Information from the inspection is more important. Our goal is to increase awareness about the environmental law.

When a company is in a difficult situation you cannot push them too much. The government of Vietnam has a policy to work with companies to solve problems.

Things are very sensitive with companies like Tae Kwang [the Nike factory] right now. Companies with operations in Indonesia can move their production there if they have a problem in Vietnam. . . . We understand that, they don't need to say it . . . The province is very concerned now about the economic downturn in Asia. Ten thousand workers have been laid off this year. Investment is down. Growth is slowing. We don't want to make conditions difficult for foreign investors. We have to be even more careful about environmental regulation now. We don't want to hurt companies or scare them off.

The DOSTE and other local government agencies also sometimes serve to protect industry from regulation. For instance, as the director of one provincial department explained (personal interview—6/11/97), "Tan Mai will not be shut down even though they lose money. Tan Mai is currently losing about 25 billion dong [over $2 million] per year. So the government is not even getting any taxes from Tan Mai. They have not received over 12 billion in taxes over the last two years." Despite these losses, the factory receives special protections, and as one local level of-ficial charged, "Although people complain, DOSTE gave Tan Mai a cer-tificate saying they were in compliance, so the *phuong* and the community can do nothing. The *phuong* cannot complain or go against the DOSTE, even though we know it is wrong" (Mr. Dung—6/6/97).

The DOSTE is thus forced to try to balance pressures from below with pressures from above, and occasionally pressures from overseas. Essen-tially, they must prove they are serious about environmental protection without hurting industrial interests. This leads to a process of negotiated regulation, and responses to the most vocal community complaints.

5.2.3 Hanoi DOSTE

The Hanoi DOSTE is a relatively powerful agency with fairly large inspec-tion and environmental management divisions. The city established its own environmental standards in 1991. However, as a number of analysts have noted, these standards cannot be effectively implemented as they cover too many parameters for the DOSTE to monitor. The Hanoi standards cover ninety-five ambient air pollutants, 117 ambient water pollutants, 140 workplace air pollutants, fifteen types of particulate contaminants, and noise. Hanoi has also developed programs to move polluting factories out of its city center. To date, the city has relocated six highly polluting factories. The Hanoi DOSTE is also the first city or provincial environmental agency to negotiate an aid project directly with a foreign government. Hanoi DOSTE is now supported by the Vietnam–Canada Environment Program and the Japan International Cooperation Agency (JICA).

Dr. Lam, the director of the environmental management division in DOSTE explained the overall capacities and responsibilities of his division (personal interview—3/28/97):

The Environmental Management Division (EMD) in the DOSTE has fourteen staff, which is not enough to cover our responsibilities. The staff lacks special training on environmental issues. We also don't get enough support from local-level People's Committees.

The community is now more and more aware of environmental issues, but still at varying levels. We get four to five complaints per day at EMD. Complaints usually go to the DOSTE inspection division and we work together with them to evaluate the complaint. We respond to every complaint with either a letter or an inspection.

In the case of compensation, we go out to the factory to negotiate the compensation rate with the local people. In one case, the farmers asked for 300–400 million dong for rice that was destroyed. We ended up getting the factory to pay 50–60 million dong. We had to evaluate the real damage and arbitrate between the factory and the farmers.

We have not closed any factories in Hanoi yet because of concerns about unemployment impacts. We threaten them, but we haven't closed any. Some factories laugh at our threats. The "8th March" Textiles factory said, "Just close us down." They aren't afraid of the threat. They know we won't close them. Factories with profits may take the threat more seriously, because we know they have money to make changes. Market conditions make a difference. If they have a market for products, they care more about the threat of closure.

Dr. Khien, the director of Hanoi DOSTE, was more positive about the DOSTE's capacity and future directions (personal interview—12/10/98):

DOSTE's capacity has improved a lot. Training programs have helped us strengthen our staff. Coordination and cooperation with other city departments has also improved. Now we invite many people from other agencies to participate in our projects, and we have close relations with other ministries. We try to share the benefits of our projects with other provinces in the north.

We need an integrated strategy for dealing with pollution in Hanoi, including thinking about employment, economics, and environment. We need to enforce regulations, but to give companies a certain period to improve their performance. We need to develop new policies that create tax benefits for cleaner production investments, and economic incentives for enforcement. And if all else fails, we should relocate or close down polluting factories.

The case of the Ba Nhat chemical factory and DOSTE's efforts to relocate the plant provide an interesting example of the challenges faced by the DOSTE. As noted in chapter 3, DOSTE faced a series of battles to win approval for relocating the factory. First, the DOSTE had to contend with other city offices (such as the Department of Industry) who wanted to allow Ba Nhat to stay in its current location and continue to produce. When the DOSTE finally won approval to move the factory out of the city center, they then had to begin the process of working with suburban and rural government officials and community members to convince them to accept the factory. These efforts were blocked twice by communities opposed to the pollution before a rural community with an existing chemical plant finally agreed to accept the plant.

As Mrs. Nhu, the Vice-Director of the EMD explained (personal interview—12/26/98):

Pollution was the key issue on motivating the move. There were many complaints from the public. And the National Assembly representative from Hai Ba Trung [the neighborhood of the factory] worked to push forward the decision. Ba Nhat is the first factory in Hanoi to be moved by force because of public pressure.

We had a multisectoral meeting to make the final decision. The Departments of Industry, DOSTE, Department of Planning and Investment, People's Council, Hai Ba Trung district, and Communist Party all participated. The Hanoi People's Committee made the final decision.

It was a long and complicated discussion. Two communities rejected the relocation of the factory to their locales. This made it a big headache for the city. It is very complicated to coordinate moving factories. We need a focal point department, and to integrate it with land use planning.

Despite institutional weaknesses and interagency conflicts, the Hanoi DOSTE is learning how to enforce environmental regulations even against city-owned factories. The agency has begun to leverage community complaints to strengthen its own position in interagency disputes. This is critical as pollution incidents are still handled through negotiation and political maneuvering between government agencies.

5.3 Pathways to Environmental Enforcement

Under the current legal and institutional framework in Vietnam, there are numerous formal incentives, but few actual pressures for firms to comply with environmental laws. Fines are too low to motivate changes. The threat of closure is nonexistent for any but the smallest firms. EIAs are largely ignored. Monitoring and inspections are still all too rare. And even when inspectors do show up, they are often ill-prepared and underpaid so they can easily be bought off or blocked.

However, regulation and punishments do sometimes occur under essentially a complaint-based environmental protection system. In this system, the more a community complains, the higher priority a problem receives from government inspectors. Public complaints appear to be the primary means to motivate state agencies to pressure firms to reduce pollution. Because agency capacity is so weak, virtually the only time these agencies go out and actually enforce is when they are pressured by community members. Ironically, this public pressure also strengthens the agencies, giving them cover to regulate, and supporting their own requests for more resources from the state.

Obviously, there are potential weaknesses and problems with this system of complaint-driven inspections and compensations. Weak capacity and poor coordination limit the scope and depth of regulation. Compensations never fully pay the costs of pollution. State officials still generally side with the polluters. Community members only focus on certain types of pollution problems. Nonetheless, these complaints can lead to real environmental improvements.

A somewhat unusual feature of Vietnamese environmental institutions is that the provincial DOSTEs have very weak formal ties to the MOSTE or the NEA. This is particularly unusual because all of the other line ministries have formal relations with the equivalent departments at the provincial level (e.g., the Ministry of Agriculture to a provincial department of agriculture). As it turns out, this is actually a critical institutional factor for environmental policy implementation in Vietnam. There are of course problems with this separation of powers and responsibilities, and problems of coordination and communication. For instance, the NEA develops and advances national policies on environmental standards, EIAs, inspections, fines, and so forth, and the provincial DOSTEs are supposed to implement them. Often the policies do not match the local "concrete conditions" in the provinces, and so the policies are not enforced. Without any formal reporting or responsibilities to NEA, the DOSTEs also feel fairly free to ignore NEA policy pronouncements and recommendations. And it appears that the only time the DOSTEs really coordinate with the NEA is when the central government (NEA or MOSTE) pay the DOSTEs to do inspections or compile "State of the Environment" reports. Numerous international reports on Vietnam's environmental institutions point to this problem and recommend stronger formal links between the NEA and the DOSTEs (World Bank 1994, UNDP 1995, World Bank 1997, UNIDO 1998).

However, there is also a potential benefit to this institutional arrangement. Local communities seem to have been able to advance their interests for better environmental protection within these confines. There are several reasons for this. First, because the DOSTEs report directly to provincial People's Committees, which report directly to the Office of the Government and the Prime Minister, complaints that make it to the DOSTE and People's Committee, can have a much greater effect on government action than if they went through the MOSTE bureaucracy on their way to top decision makers.

Second, while MOSTE, and now MONRE, is in theory equal to other ministries, it is in fact much weaker and generally loses debates in the Council of Ministers that affect the interests of the Ministry of Planning and Investment (such as foreign investors) or the Ministry of Industry (such as state-owned enterprises). MOSTE is thus rarely able to advance environmental concerns through the political hierarchy. As I have noted, the NEA and MOSTE are not powerful enough to have a factory closed. However, provincial and city DOSTEs have been successful in getting factories shut down or moved. This seems to be due to the fact that the NEA and MOSTE have more potential roadblocks on their paths to policy implementation.

Third, the NEA and MOSTE are much less responsive to public complaints and concerns than are local officials. Sitting in Hanoi, the NEA and MOSTE are largely insulated from local political pressures. DOSTEs are thus better "targets" for local complaints. Local community members can complain directly to local officials, go to their offices, write letters to officials and local newspapers, and force some accountability on government agencies. So although the DOSTEs may have less capacity and official power than the NEA, because of their responsiveness to local demands, and their connections to provincial decision makers, they are more likely to enforce environmental regulations than national officials.

It is interesting to note how often pollution issues do attract the attention of the highest levels of government based on community complaints. In three of the case study firms (Tan Mai, Lam Thao, and Viet Tri) the Prime Minister or Vice-Prime Minister visited the plants and made public statements about solving their pollution problems. In the Ba Nhat case, community members wrote letters to the Prime Minister and to members of the National Assembly, motivating the Assembly to discuss the case during a session. As the Economist magazine has noted, mishandled complaints "have prompted the prime minister, Phan Van Khai, to dispatch five high-level delegations to the fifteen provinces to hear popular complaints and deal with them. The delegations, some of which include deputy ministers, have been told to check that local authorities are handling complaints properly, and to prod local authorities into resolving difficult cases" (EUI 2000b: 16).

The ability to actually impose sanctions on firms may also be affected by the interactions between the promoting agency and the regulating agency. For instance, if the DOSTE is up against the Ministry of Industry it will have different success than when it is up against a locally owned factory or a foreign firm. Different firms will have different connections to the state that can protect or insulate them from environmental regulation. This "match-up" will also influence the effectiveness of public complaints.

Processes of community participation and complaints can also benefit state environmental agencies in a number of ways. First, by transmitting information on environmental problems, community members can serve as dispersed monitors, and prioritizers of environmental concerns. Complaints signal which issues are of primary concern to citizens and what minimum actions are needed to prove the state is doing its job. Community complaints can also be used by savvy environmental officials as ammunition in their own campaigns to increase their power, resources, and standing. As Roodman (1999b) explains, "when an inspector visits a plant to press for changes, knowing that neighbors had complained about the plant made his or her job easier."

5.4 Legitimacy and State Action

Since at least the 1940s, the Vietnamese Communist Party has understood that its main source of support and strength was rooted in the provision of economic and social benefits to the peasant class. The foot-soldiers of early liberation struggles were rural peasants. Land reforms instituted in the 1950s were the cornerstone of Communist Party efforts to mobilize peasants in the North. As Kolko (1995: 33) notes "the Party's success as a social movement was based largely on its responsiveness to peasant desires." Even the recent transition process has been heavily influenced by peasant demands. Fforde and de Vylder (1996: 15) assert that "'Reform' in Vietnam was largely a bottom-up process. . . . 'Policy' was, therefore, responsive, rather than proactive."

It is said in Vietnam that "The writ of the king bows to the customs of the village" (*"phep vua thua le lang"*) or as Fforde and de Vylder (1996: 50) explain, "local autonomy was part of the accepted balance of power between the central Imperial Court and the local communes." Despite decades of central planning, a surprising amount of local autonomy still exists in Vietnam which can be attributed in part to the need for decentralized action during years of war.

This has led to a wide range of development decisions being contested in Vietnam. For example, peasants in Thai Binh province have forcefully opposed local taxation policies and corrupt government officials. Women in Dong Nai province have opposed government efforts to reallocate land. Farmers just outside Hanoi have violently struggled to protect their land from a government-supported golf course development. And most recently, villagers near Qui Nhon, in central Vietnam, actually stormed a factory, tore down its walls, and destroyed its equipment to protest local pollution (Associated Press 2002).

One component of recent public concern regarding development has focused on the environmental impacts of industrialization. Since 1994, complaints and protests around environmental issues appear to have grown exponentially.[2] With the passage of the Law on Environmental Protection, the public has made clear that it is not willing to bear the environmental and social costs of industrial development while others profit from this development. As Beresford and Fraser (1992: 15) argued even before the passage of the LEP, "What is increasingly being called into question is the ability of the traditional socialist model to deliver both economic growth . . . and development broadly interpreted to include rising living standards, improved health, more leisure and more democracy for the majority of the population in whose name the socialist revolution was carried out. Containment and prevention of environmental damage is increasingly being seen as one of the necessary conditions for this goal."

As with other issues, the state has responded to public demands with a careful mixture of both rhetoric and action. For instance, in the 1960s, Ho Chi Minh declared that the "Forest is gold. If we know how to conserve and use it well, it will be very precious" (cited in Buffet 1996). General Vo Nguyen Giap proclaimed twenty years later, "The soldier comes to another front now, the environmental front. . . . without environmental recovery, Vietnam cannot have economic recovery" (cited in Beresford and Fraser 1992).

The Vietnamese state appears to be in an interesting position regarding environmental issues. By staking its claims to legitimacy on its ability to chart a development course that promotes economic growth while mitigating the adverse social and environmental impacts of unbridled capitalist industrialization, the state comes face to face with a number of internal contradictions. For instance, the state must both attract foreign investors, and at the same time attempt to regulate these actors. The state must balance its support for the emergence of a capitalist class, with the need to continue protecting the interests of peasants and workers.

As Gabriel Kolko (1995: 33) warns, "a Party that triumphed because of the social role and needs of the peasantry, and then alienates them, is risking suicide." The Party is thus charting a careful course that involves responding to certain community demands as a means to balance (or pacify those affected by) adverse impacts of capitalist economic development. Demands for environmental quality are accepted as one trade-off in the transition to the market.

In this regard, the state's struggle for legitimacy, and local government sensitivity to accusations of corruption, have opened up new space for community action around environmental issues. Vietnam's socialist legacy has essentially provided a window of opportunity for community participation, and civil society more broadly, to play a constructive role in pollution issues. And through this small window, many Vietnamese community members have jumped.

One might reasonably ask why the environment has become an open issue for debate while so many other issues remain off limits? Why for instance are journalists able to write about state owned enterprise pollution and ineffective state regulation, but not about petty corruption in a customs office (which recently landed the editor of a major newspaper in jail for exposing "state secrets")? Here again I believe the issue of state legitimacy and the government seeking to find a balance between community concerns and development policies has been critical to opening space for discussion about uses of the environment.

The Vietnamese government has also felt external pressures from foreign donors and even the World Bank to establish environmental laws and institutions. The government signed all the major conventions at the Rio summit on sustainable development in 1992 and has been in line ever since for Capacity 21 funds from the United Nations to implement sustainable development programs and policies. These external agreements and pressures have helped push the government to pass laws and to tolerate (if not encourage) public participation. Some have argued that this external pressure is a new form of green colonialism (Goldman 2000). Nonetheless, in the Vietnamese case, it has seemed to promote regulatory action.

Vietnamese citizens now literally file thousands of complaints against industrial polluters each year, and journalists produce hundreds of stories (Roodman 1999a). Ho Chi Minh City's DOSTE reports receiving over 1,000 complaints per year. The Hanoi DOSTE similarly receives around 1,000 complaints per year, while Dong Nai receives approximately 200 complaints per year. The DOSTEs also appear particularly sensitive to media criticisms of their failure to enforce pollution laws. Perhaps it is because these critiques conjure an accusation of corruption and incompetence that they usually respond so strongly to media reports.

The passage of the Law on Environmental Protection was a critical step in creating a legal opening for community complaints and demands. Article 43 of the LEP states that "Organizations, individuals have the right to complain, denounce to the State management agency for environmental protection or other competent State agencies about activities in breach of environmental protection legislation." As Roodman (1999b) points out, this right of "complaints from people living near factories, along with media pressure, have perhaps played at least as significant role as conventional regulatory measures in driving industrial pollution reductions in Vietnam."

This opening and the right of citizens to criticize the government does of course have limitations. Community members allude to unstated lines that must not be crossed. For example, people generally avoid publicly accusing high-level officials of corruption. And the Law on Environmental Protection itself is quite restrictive in how the public can participate in environmental decisions, essentially only allowing complaints after pollution has occurred.

5.5 Conclusions

The Vietnamese state has been undergoing major institutional and administrative changes over the last ten years. And although the Communist Party continues to affirm its sole control over the government, the roles and capacities of the state are nonetheless being transformed. Most importantly for this study, the state now has less control over production and economic decisions, and has been forced to develop new bases of legitimacy and public support, and new mechanisms for dealing with industrial polluters. Evolving environmental policies and implementation practices must be viewed against this backdrop of institutional changes.

Based on recent high level policy statements, it appears that the government recognizes a number of weaknesses and problems in its environmental protection efforts. The government is thus working to shore up its environmental agencies and authority, and to move gradually toward more effective protection policies. However, this is a long and slow road.

At present, environmental agencies at all levels in Vietnam have very limited capacities for environmental regulation. They also have limited political powers in internal government battles and day-to-day negotiations with other government agencies. Corruption among poorly paid inspectors only adds to the challenges of regulation.

However, public and media pressures on environmental issues are gradually raising the profile and the bargaining power of environmental agencies. Community pressures have helped to overcome agency resistance to implement laws that impact other state actors (such as the Ministry of Industry). Community pressures are also motivating inspectors to simply do their jobs, a not insignificant feat as most inspectors are overwhelmed by their tasks, undertrained for their duties, and underpaid. Community action also helps shine light on local-level corruption, and increases transparency in state environmental actions.

Vietnam now thus has essentially a complaint-driven environmental protection system. Both inspections and sanctions are motivated largely by community demands. The state obviously has played an important role in creating the legal structure for this system, and, perhaps accidentally, for creating the institutional arrangements that make the DOSTEs an effective target for public pressure. To its credit, though, these state agencies have worked hard since they were established in 1995 to build their own capacity to gather environmental information and to monitor compliance with the country's new laws and regulations.

As the interviews with DOSTE staff made clear, agency relations with factory managers are often less rigid than described in laws and regulations. Seldom in fact, are regulations applied along the letter of the law. And it is virtually unheard of in Vietnam for lawyers to participate in legal analysis of statutes, their implementation, or firm compliance. Where litigation is the name of the game in countries such as the United States, negotiation is the critical process in Vietnam. Virtually every aspect of environmental policy and regulation is open to negotiation, including fines, compensations, relocations, and timelines for remediation.

The staff of the DOSTEs are acutely aware of their limited authority, and of the social and political pressures against fining or closing factories. Agencies thus seldom take the political risk of going out on their own and enforcing the law. In fact, it is often only after they receive strong or repeated complaints from the public that they feel legitimated in their duties. Even then, DOSTEs attempt to mediate and negotiate with affected parties rather than enforce rigid standards and sanctions.

The Center for Environmental Management's system of negotiation is an excellent example of this. The director of CEM explained several times how his job was to evaluate pollution incidents and then figure out what a company could afford to pay without causing job losses or other undue economic impacts. Decree 26 and Decree 175 were backdrops to these negotiations, but by no means did they take precedence or mandate final decisions.

In this system, malfeasance is essentially determined by evaluating a firm's attitude and intent. Regulators ask whether the firm meant to pollute, whether they had tried to avoid it, and whether they had taken measures to fix their problems. Firms that are "trying" to comply are treated more leniently than those that appear to be flaunting the law.

On the ground, when pressured to act, the DOSTEs work to deliberate and negotiate over environmental enforcement. Sanctions, compensation, and compliance schedules are all negotiated. This process of complaint-driven, negotiated regulation looks much different in practice than legal mandates for environmental regulation appear on paper. So, while the official legal structures are largely ineffective, some community and state actors have hit upon alternative mechanisms and pathways to enforcement.

This process seems to work because of the institutional structure of DOSTE reporting and responsibility, and because of the DOSTE's sensitivity to public criticisms. Media reports and public letter-writing campaigns raise the implication of corruption or incompetence within the DOSTEs. This pressure from below, combined with the pressure from the hierarchy above, which reaches all the way to the Prime Minister and the Office of the Government, can serve to squeeze the DOSTEs into action. At the end of the day, the DOSTEs must respond to both competency concerns and larger questions of state legitimacy.

The case studies seem to point to a number of general conclusions about the role of the state in this complaint-driven process. First, state agencies need a certain level of autonomy from industrial actors, otherwise economic interests take precedence over environmental concerns. If the state is too closely connected to a firm, regulation becomes virtually impossible. This can occur in both state enterprises and joint ventures. Second, state agencies need a certain level of connectedness with the communities affected by the pollution. There needs to be some form or process of communication for communities to successfully impart their complaints and goals. Here again, this is why the DOSTEs rather than the NEA seem to be so much more effective in environmental enforcement.

As I have mentioned, community complaints and organized pressures are a critical driver of this process. It should be noted however, that the state is by no means always happy to respond to community demands for stricter environmental protection. Despite positive examples, state agencies do regularly attempt to block community action. Essentially all protests in Vietnam are illegal, and most state enterprises appear to exist above the law. The range of community actions is thus circumscribed by state decrees and the unwritten rules protecting state enterprises. Vague legal rights create barriers for communities to complain, seek compensation, and demand action against a firm. State agencies also sometimes intervene on behalf of state enterprises and joint-ventures where the state has a direct financial interest in the factory.

However, if state agencies can be motivated to respond to community demands, this complaint-driven process can result in effective regulation, and in turn boost the capacity and effectiveness of local environmental regulators. In several of the case studies, state agencies played pivotal roles in supporting and legitimating community demands for pollution reduction. Successful community actions responded to these openings and focused pressure on the right actors within the state.

6

Information, Accountability, and Direct Pressure Politics: The Role of Extralocal Actors

Community dynamics, like most politics, are local. However, the globalization of production and consumption is increasingly expanding the scope of impacted communities, and creating transnational networks of interested stakeholders. Perhaps no company in the world has symbolized this globalization of interests and concerns as Nike, Inc., the world's leading footwear company.

To give an example, in early 1998, a worker was allegedly hit by a manager in one of Nike's factories in Vietnam. A friend of this worker called Thuyen Nguyen, the Vietnamese-American director of Vietnam Labor Watch in Washington, DC, to report the incident. Mr. Nguyen had given this person a prepaid phone card to use in such occasions. After learning about the abuse, Mr. Thuyen called another contact in Vietnam to arrange for the worker to be interviewed on videotape the next day. This videotape was then sent to the Singapore office of ESPN, the cable sports channel, which was preparing a story on conditions in Nike factories in Vietnam. The interview was transmitted to ESPN headquarters in the United States and translated into English. The tape and transcript were also circulated to interested newspaper reporters. Within a matter of days, the interview with the abused worker appeared on televisions around the world, before Nike officials in the United States had even heard about the incident.[1]

This example around Nike is obviously quite different from the previous cases of direct local community participation in environmental disputes. Although located in Vietnam (and China, Indonesia, Thailand, and other countries), Nike factories are actually the focus of transnational communities of stakeholders. These networks include international NGOs, local NGOs, workers, and the media engaging companies such as Nike—

globally branded firms with local operations—and seeking to both change local conditions and to transform global systems of production.

Understanding how these international networks operate in a country like Vietnam, and how they connect to local nongovernmental organizations, is complicated. The lines between the state and civil society alone are sometimes hard to distinguish in Vietnam. For example, as part of my research I interviewed the director of the Department of Science, Technology, and Environment (DOSTE) of Hai Phong province. I arrived at the government office building and was led down a long hall past a series of People's Committee offices. Once inside the office with the interview underway, I was confused as the man leading the meeting kept referring to the Environmental Impact Assessment (EIA) reports his nongovernmental organization (NGO) had completed. After a while I stopped him and asked whether I was in the DOSTE or the office of an NGO. He smiled and said "Both. We wear two hats here, a DOSTE hat, and an NGO hat. Sometimes I am the head of the DOSTE, and sometimes I am the head of my NGO." As he explained proudly, his NGO was involved in writing EIAs for foreign factories locating in the province, and then his DOSTE was responsible for reviewing these EIAs.

This experience reinforced the sense that NGOs in many developing countries are sometimes "just a bureaucrat with his own letterhead" as Ferguson (1998) notes for Africa. In Vietnam, this is particularly complicated as "nonstate" actors can represent very different constituents. In fact, the penetration of the state into civil society in Vietnam makes it difficult sometimes to differentiate between state and nonstate actors. And while these groups seldom resemble western NGOs, they often can play critical roles in environmental debates.

The role of nonstate actors is particularly critical when communities come up against powers they simply cannot influence. Groups interested in expanding industrial activity or protecting existing firms can work to block environmental initiatives. Local government agencies are sometimes unresponsive to local pressures. And corruption at the local level can stymie even the best-organized efforts to pressure the state or a firm. Perhaps more importantly, when decisions about production and pollution are made extralocally, it is extremely difficult, and sometimes impossible, to influence these decisions at the local level. Large state enterprises are managed from Hanoi. Foreign multinationals are managed from far-off countries. And subcontractors of multinational firms often get directions from regional centers. Community efforts thus sometimes require external resources, political support, or connections to find and then pressure key decision-makers. But who provides these resources and what do they actually accomplish? I focus on three strategies of nonstate actors in this chapter: information politics, accountability politics, and direct pressures.

Information politics, as Keck and Sikkink (1998: 19) explain, involves at its base, "promoting change by reporting facts." Simply by gathering and disseminating information, NGOs, academics, and the media can play an important role in providing a voice to local demands, and in breaking state monopolies over information. These organizations can also help to amplify local voices, giving them more credibility and legitimacy, and to reframe debates about environmental issues. The most sophisticated NGOs work very closely with the media to advance these goals. The Nike case shows all of these processes quite clearly.

The second key dynamic involves accountability politics which seek to "expose the distance between discourse and practice" (Keck and Sikkink 1998) of state agencies and firms. NGOs and the media can raise issues of agency effectiveness and commitment, or firm compliance with laws and codes of conduct. These strategies help broaden pressures by exposing irresponsible actions or hypocrisies to the public at large, thereby creating greater pressure on state agencies to resolve problems. International campaigns have been particularly effective in targeting brand-sensitive firms—firms that spend large amounts of money advancing their brand image—and pointing out hypocrisies or contradictions in their marketing.

Third, NGOs and other extralocal actors can also employ direct pressure politics. International NGOs organize letter-writing campaigns, protests, and boycotts, which demand specific changes in firm practices. NGOs, academics, and the media also use their influence to redirect accountability onto state agencies or corporate offices that are not targeted by community members. In this way pressure can be applied to different decision makers in different ways.

This chapter examines these emerging networks of nongovernmental actors in Vietnam, both local and extralocal, and their efforts to advance information, accountability, and direct pressure politics. Cases in Vietnam provide new insights into how these transnational networks operate, how they connect to local actors, and how they can influence state and corporate actors.

6.1 The Nike Case—Tae Kwang Vina

In a single room the size of a football field, 2,000 women sit hunched over sewing machines stitching sports shoes. Row after row of young women work eleven hours per day, six days per week at the production lines of the Korean-owned Tae Kwang Vina company.[2] Most of the factory's workers (90 percent of whom are women) have traveled from northern and central Vietnam in search of these jobs. Trading rural rice fields for the new Bien Hoa 2 Industrial Estate, the women have left their homes and families for the prospect of forty dollars per month and a better life.

The vision of a better life is hard to conjure walking through this factory of approximately 10,000 workers. During the summer months the workers sweat in hundred-degree-plus temperatures at their assembly lines. Figure 6.1 shows women assembling shoes in the plant. When I first visited the factory, the workers were exposed to toxic solvents and glues that often made them dizzy or nauseated (and that would be much more strictly regulated in countries such as the United States). Respiratory ailments were common, as were accidents in some of the more hazardous sections. And then there was the repeated verbal and physical abuse they experienced at the hands of foreign managers.

More than 7,000 miles away, in cities such as Portland, San Francisco, and New York, human rights and labor activists have been strategizing on how to change the conditions inside factories like Tae Kwang. Although the activists might not know the name of this factory, who its managers are, or who even owns it, they are clear on who is responsible for its poor labor and environmental conditions: Nike, Inc.

With $9 billion in sales in 2000, Nike is the world's leading producer of sports shoes and apparel. It is also one of the world's leading innovators in global outsourcing. Nike owns none of the factories that produce its famous sports shoes. The five Nike factories in Vietnam, which employ 35,000 workers, are owned by Korean and Taiwanese subcontractors. Nike still designs its shoes in Beaverton, Oregon, but prototype shoes are pro-

Figure 6.1
On the assembly line at Tae Kwang Vina

duced in Seoul or Taipei, and a final production run is likely to be done in China, Indonesia, or Vietnam. For twenty years, this subcontracting arrangement was a win-win situation for Nike. The company was able to create competition between subcontractors, push down production costs, shift the risk of investing in factories and equipment, and avoid the difficulties of managing hundreds of thousands of workers. It was also able to use the subcontracting system as an excuse to avoid responsibility for environmental and working conditions in the factories that produce its shoes. As Nike neither owned nor managed the factories, it argued it could not be held responsible for their day-to-day conditions, despite the fact that Nike staff are in the factories every day monitoring what is produced, how it is produced, and the quality of the final products.

Although Tae Kwang Vina is one of the newer factories in Dong Nai province, by 1997, it was already known as a bad place to work. In general, people from Dong Nai avoid jobs at Tae Kwang if they can help it. Dong Nai is close to Ho Chi Minh City, so finding other jobs is a viable option. Huge numbers of migrants thus serve as the labor pool for the factory. Tae Kwang is also known by government agencies as "uncooperative." Officials in the Dong Nai DOSTE say they get little response from Tae Kwang to their requests that it reduce its toxic emissions or clean up its polluted effluent. The factory seems to have learned a lot about navigating local labor laws and environmental regulations from its experiences in South Korea and China. For instance, the company has been able to avoid complying with national requirements for a wastewater treatment plant without suffering any consequences. As mentioned, it hired the son of the chairman of the Dong Nai Communist Party to help them with these kinds of issues and to be a "problem solver" with the government.

Regulating companies like Tae Kwang is obviously difficult for the Vietnamese government. The government is in the bind of working to attract foreign capital (competing with countries such as China and Indonesia, where the bulk of Nike production occurs) while at the same time attempting to establish regulatory policies and mechanisms of enforcement. Clearly, a company responsible for more than 35,000 jobs and fully 4 percent of the value of Vietnam's total exports in 1998 carries significant influence with government officials.

Environmental laws are one example of policies that have been selectively enforced in Vietnam. For instance, Tae Kwang burned all of the scrap rubber from its production process in order to generate steam, and in the process created thick, black clouds of pollution. Despite requests by the NEA and DOSTE to reduce these emissions, in 1997, Tae Kwang actually purchased additional scrap rubber from other Nike plants in Vietnam to feed its boilers, thereby increasing the pollution. Figure 6.2 shows one worker in the boiler room burning scrap shoe materials. Because Tae Kwang is in the middle of an industrial estate, no one officially lives near the factory, as the area is not zoned for housing, so there is no recognized community to complain about these problems. The people most affected by Tae Kwang's pollution—the workers—have little power to influence the company. The company union is controlled by the managers, who handpick union representatives. The community that does exist around Tae Kwang has little capacity, cohesion, or linkage to external actors. The majority of the factory's workers are recent immigrants to the province, often straight off farms, who rarely stay more than a year or two at the fac-

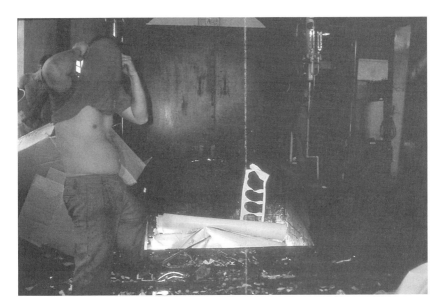

Figure 6.2
Burning scrap materials at Tae Kwang Vina

tory before moving on. The residents of the area appear to feel few connections or allegiances to these workers.

Local efforts by either government officials or community members to
influence Tae Kwang have thus had little effect. However, powerful external pressures do appear to have influenced the company. NGOs in the
United States and Europe have been successful in pressuring Tae Kwang to
change its production practices and have helped to build the capacity and
linkages available to workers and community members. Activist campaigns
regarding labor conditions in Nike plants have gained worldwide media
attention. In October 1997, groups in more than ten countries organized
protests, pickets, and informational campaigns regarding Nike's production practices. NGOs in the United States (such as Global Exchange, the
Campaign for Labor Rights, Vietnam Labor Watch, Press for Change, Justice Do It! Nike, Sweatshop Watch, the Transnational Resource and Action
Center, and the National Labor Committee) and in Europe and Australia
(such as the Clean Clothes Campaign and the NikeWatch project) have
coordinated major campaigns. And NGOs in Hong Kong, Indonesia,
Thailand, Cambodia, and Mexico have supported and helped organize

workers on the ground against Nike. These groups regularly conduct research on factory conditions, interview workers, and publish reports on
the Internet. These groups have also used the media to educate the public about conditions inside Nike plants. Some have called for boycotts of
Nike products, while others have begun lobbying government bodies to
require Nike to change (Benjamin 1999, Bissell 1999, Shaw 1999). My
own research on Tae Kwang has been used by these groups to pressure
Nike to improve conditions in its Vietnamese plants (Greenhouse 1997,
O'Rourke 1997).

For Tae Kwang, extralocal pressures have led to regular visits from
Nike's labor and environmental inspectors, as well as monitoring by third-
party auditing firms. The first goal of these inspections has been to reduce
worker exposures to toxic solvents and glues. In May 1998, the company
announced a major initiative to eliminate the use of organic solvent-based
cleaners and glues, pledging to comply with U.S. workplace laws in all of
its factories. By December 1998, workplace health and safety conditions
had been substantially improved at Tae Kwang (O'Rourke and Brown
1999). Pressure from Nike—which has been driven by media and NGO
pressures in consumer markets—is now having more impact on the reduction of air pollution and workplace hazards than local government or
community pressures could have had on their own.

The Tae Kwang case is obviously quite different from the other five cases
examined in this book. There are no neighbors living next to the factory,
and the workers have no cohesive community to represent their environmental and health concerns. Instead, a network of NGOs and activists
outside Vietnam has worked to pressure Nike directly and to indirectly
pressure the Vietnamese government to increase enforcement. This coalition of international actors has then worked to build links to local community members and workers. In this way, extralocal actors are actually
working to support and build the capacity of local community members
and are providing room for state agencies to more effectively regulate a
multinational corporation.

Both the foreign press and international NGOs have publicized problems in Nike's factories in Vietnam. This information has played a critical
role in the transnational advocacy networks that connect workers in developing countries, to local NGOs, developed country NGOs, unions, the

media, and consumers. International campaigns are now closely coordinated with local campaigns, linking trade union pressures on factories such as Tae Kwang to consumer pressures on merchandisers such as Nike, Reebok, and adidas. These information-based, transnational campaigns are now pushing for fuller public disclosure and independent monitoring of factory conditions around the world, and are working to advance corporate codes of conduct that are stronger in practice than local laws. All of this seeks to connect back into processes in Vietnam (and other developing countries) to influence government officials and workers themselves.

For example, while interviewing factory workers one night in Dong Nai, a worker from Tae Kwang sounded out a word in English that I initially could not understand. Several times he said "cotboy," "cotboy." I asked him to explain the word in Vietnamese, at which point I realized he had been trying to say "Boycott!" It is surprising enough that he knew of calls to boycott Nike, which is not even widely known in the United States. But it is even more surprising that he would support international actions that could potentially cost him his job. In many ways his knowledge of the boycott is a testament to advances in communication technologies and networking strategies. In the Nike case, local concerns about factory conditions and impacts have been significantly informed and influenced by international campaigns. Information politics, accountability politics, and direct pressures are all at work in this case.

At the same time, it is almost equally surprising that people in the United States are concerned about health and environmental impacts at factories in Vietnam. To be sure, no one in the United States has probably ever heard of these case study firms. Even people who follow the Nike debate usually do not know the names of the actual factories in Asia producing these shoes. Nonetheless, Tae Kwang has become a high profile corporation in Vietnam, and internationally is now held up as a poster boy for global sweatshop issues. This would never have happened if Tae Kwang produced generic shoes. Being associated with Nike, the world's leading marketer of sports shoes, has catapulted the Korean contractor into public prominence. An unrelenting focus by NGOs and the media has kept Tae Kwang and other Nike factories in the spotlight.

The Tae Kwang case actually has a quite limited local community role. Tae Kwang has effectively insulated itself from community (and worker) pressures through a number of strategies. By siting in an industrial zone, Tae Kwang has reduced potential pressures from neighbors. Air pollution from Tae Kwang's burning of scrap rubber, nylon, and plastics primarily affects factories sited nearby (who regularly complain to the DOSTE). Workers are essentially imported to the factory—which is in the middle of an industrial park—on bikes or buses every day. And over 70 percent of workers at Tae Kwang are recent migrants to Dong Nai province from central and northern Vietnam, so they do not have attachments to the local community. Because of this separation from the local community, advocacy networks are even more important.

6.2 Nonstate Actors in Vietnam

There are a range of nonstate actors and informal networks that do operate within Vietnam. There are now over 150 environmental centers, institutes, and organizations registered in Vietnam. The most prominent of these all call themselves NGOs.[3] However, while they operate like NGOs, these groups are often associated with government universities or institutes. These organizations have autonomy over individual projects, assuming they can raise funding, but they are constrained in what they can critique or recommend. All of these groups are careful to avoid being overly critical of government policies or programs.

Over the last five years, these groups have gained increased autonomy from the state. Their staff travel more frequently overseas. They regularly work with foreign organizations and seek funding from foreign donors. And they are building up the expertise and resources to expand their services. Nonetheless, they must still respond to state requests first and foremost, and because of their limited funding they continue to operate more like consulting firms than advocacy organizations.

In Vietnam, NGOs and other nonstate actors operate differently than similar organizations in the United States or Europe. Although the Communist Party has sought to enter into all realms of civil society, independent, critical actions do occur in a space between the individual and the state. Mass organizations (such as the Women's Union, Peasant Union, Youth Union, and Fatherland Front), as well as academics, academic NGOs, and the media all operate to different degrees like advocacy organizations.

International NGOs, international governmental agencies (such as UNDP), bilateral aid programs (such as the German aid agency GTZ and Sweden's aid program SIDA), and the international media also play a role in Vietnamese environmental politics.

A number of NGOs have been growing significantly over the last five years. Groups such as the Women's Union and the Vietnam General Confederation of Labor are difficult to pigeonhole as they play varied roles, and represent multiple constituencies. These groups were designed to do the political and cultural work of the party, but through their own political decisions, and with the help of international assistance, now also develop initiatives of their own. To be sure, these groups do not conduct Greenpeace-style civil disobedience, but they do play important roles in political debates, and are part of a larger political process in Vietnam.

Local NGO activities in Vietnam are often constrained in ways that transnational advocacy networks are not. Local NGOs cannot outwardly attempt to influence the government. They are forced to frame their work in broader activities such as capacity building, education programs, and women's empowerment. NGOs in Vietnam thus work to show that things can be done differently and to legitimate local voices in national debates. They also help to transfer successful models from community to community and to support social norms and networks that operate outside state control.

International NGOs have flooded into Vietnam over the last five years in lock step with renewed development assistance. There are now over 160 foreign NGOs registered in Vietnam, several dozen of which work on environment-related projects. World Wildlife Fund, IUCN, Fauna and Flora International, and Birdlife International all have ongoing programs in Vietnam. Development organizations such as Oxfam, CARE, Save the Children, and the American Friends Service Committee also sponsor environmental projects.

In the last several years, a number of these NGOs and several large U.N. and bilateral aid projects have taken a decidedly "brown" turn, that is, towards urban and industrial environmental issues. Whereas early international environmental projects focused on forest conservation and species protection (what the World Bank calls "green" projects), many recent projects are focused on pollution control (the "brown" side). At last count there were over twenty international projects operating in Vietnam focused on industrial pollution control.

Vietnam has also been invaded by an army of academic researchers and foreign journalists, including a whole generation of future professors currently cutting their teeth in the Vietnamese field. These researchers are producing a flurry of English-language reports and articles pointing out both the benefits and the costs of Vietnam's capitalist development. Environmental degradation and worker exploitation are hard to overlook in analyzing Vietnam's transition to the market. This reporting shines a previously absent spotlight on government decision making, raising issues of accountability, effectiveness, and fairness. Vietnamese researchers are also being transformed through collaboration and funding from international sources. Critical analyses of the environmental impacts of industrial development—a topic of interest to foreign funders—are much more common now than only a few years ago.

6.3 Goals and Strategies

The first strategy these advocacy networks employ is to gather information on conditions and practices inside factories or within communities. International activists have traveled to Vietnam to conduct research, to build relationships with local organizations and networks, to connect with individual workers and community members to create channels for gathering up-to-date information on conditions on the ground. International NGOs have focused on particularly egregious cases of worker abuses and pollution incidents. For instance, international NGO reports often paint a picture of life in Vietnamese factories that starkly contrast with the claims made by firms such as Nike.

Their second task is to get stories to the public in the United States and Europe, the primary consumer markets for branded shoes and apparel. International NGOs have been extremely adept at working with the media to publicize stories they uncover. NGOs pick up stories reported in the Vietnamese press, translate them into English, and circulate them to other NGOs and the media in the West. They also produce their own research reports. Reports by Vietnam Labor Watch, the National Labor Committee, and Global Exchange have received significant coverage in the U.S. media. Global Exchange had its own media relations firm which worked with reporters to get stories written and editorials placed in prominent media.

Information gathering and dissemination are coordinated so that they are extremely fast and flexible. Information literally flows at the speed of the Internet, as with the case of the worker hit by the manager. Reports are written in a form that reporters can use, with quotes from workers, and documentation to support claims. Several groups have also sponsored speaking tours of Nike workers to the United States to put a human face on working conditions and management practices. These tours have received significant press coverage and been used to organize and raise awareness of U.S. consumers and students.

The third task these groups carry out is to directly pressure firms such as Nike to change practices at subcontractor factories. International NGOs have had very limited success in influencing companies such as Tae Kwang or Pou Chen (the largest Nike subcontractor). Instead they focus on Nike, which does not actually own the factories that manufacture its shoes. NGOs in effect pressure Nike to pressure contractors like Tae Kwang. They realize the difficulty of pressuring a small Korean firm operating in Vietnam, and at the same time, recognize the potential power of protests and boycotts to Nike.

These protests and media campaigns clearly do impact the factories in Vietnam. For instance, as one Korean manager explained (Mr. Im— 3/14/97), "There were protests in front of NikeTown [in the United States], so now Nike wants us to change." And another Vietnamese manager at Tae Kwang who reported (Mr. Nguyen—6/6/97), "We went to a meeting at Nike yesterday in Ho Chi Minh City. Dr. Tien [Nike's labor practices manager] showed all the newspaper reports from around the world on conditions at Nike plants. . . . He warned us to improve our labor practices."

Because Nike is so image conscious, association with sweatshops and pollution has the potential to cause long-term damage to its brand, and ultimately to its sales. U.S.-based NGOs have employed a number of traditional pressure tactics against Nike, including protests in front of Nike-Town stores, letter writing campaigns, and sponsoring debates on college campuses. NGOs have worked with college students to get Nike products removed from campus stores, and to end Nike endorsement contracts with university sports programs. These actions continually hammer home the message that Nike is responsible for conditions in it subcontractor factories, and that Nike should solve these problems.

By attacking Nike's image (which the company spends approximately $1 billion per year promoting), NGOs in the United States seek to pressure Nike into forcing its subcontractors to change, and then to monitor the performance of these factories. Broad-based demands for companies such as Nike to enforce a code of conduct for their contractors, and then to openly monitor compliance with this code, has led to the creation of a number of new organizations to certify contractor compliance with codes. Nike has supported one such organization—the Fair Labor Association—and has agreed to allow NGO participation in monitoring its factories. Student and union groups have established their own system of monitoring and verification, called the Workers Rights Consortium, which would directly involve local NGOs and unions in evaluating conditions inside Nike's factories around the world.

The fourth task of these groups is to motivate action on the part of the Vietnamese government, and more broadly to strengthen the enforcement of labor and environmental laws by host governments. The Vietnamese media has focused on this goal as well. The *Lao Dong* (Labor) and *Thanh Nien* (Youth) newspapers have been especially critical of Nike subcontractors in Vietnam, putting indirect pressure on the government to regulate these firms more strictly. Newspapers allude to Vietnamese nationalist rhetoric, questioning whether Vietnam won the war against the Americans only to now have Vietnamese workers exploited by capitalist American companies.

As mentioned previously, I have become a participant in one information network focusing on Nike. I was brought into this network through a specific piece of information I uncovered in my research. While undoubtedly this participation has shaped my view of the overall case, it has also allowed me an unprecedented opportunity to examine how different NGOs and media actors have sought to influence a major multinational corporation operating in Vietnam and how Nike has responded.

During my research, I was leaked an internal report commissioned by Nike, and written by the international accounting firm Ernst & Young.[4] The report summarized an audit of the labor and environmental conditions inside the Tae Kwang Vina factory completed in January 1997. The report, while not methodologically or technically rigorous, was nonetheless quite critical of the workplace and environmental conditions inside Tae Kwang Vina. The audit noted violations of labor laws on wages and working hours, unprotected chemical exposures, high levels of respiratory illnesses among workers, and violations of environmental laws. The audit turned out to be the kind of smoking gun NGOs and the media had been looking for.

When I returned to the United States in October 1997 after completing a phase of my fieldwork, I made the Ernst & Young report available to the media and wrote a short report explaining my own findings inside Tae Kwang Vina (O'Rourke 1997). The report struck a cord with the U.S. media and the public. A front page *New York Times* article summarized the auditors' analysis "that workers at the factory near Ho Chi Minh City were exposed to carcinogens [sic] that exceeded local legal standards by 177 times in parts of the plant and that 77 percent of the employees suffered from respiratory problems," painting "a dismal picture of thousands of young women, most under age twenty-five, laboring 10.5 hours a day, six days a week, in excessive heat and noise and in foul air, for slightly more than $10 a week. The report also found that workers with skin or breathing problems had not been transferred to departments free of chemicals and that more than half the workers who dealt with dangerous chemicals did not wear protective masks or gloves." The *Times* concluded that "the fact that such conditions existed in one of Nike's newer plants and were given a withering assessment by Nike's own consultants made for yet another embarrassing episode in a continuing saga" (Greenhouse 1997). Business Week, the Associated Press, Reuters, Agence France Presse, the Wall Street Journal, and a number of regional papers followed up with stories of their own criticizing Nike.

NGOs also ran with the story. Environmental health and safety issues, which had previously taken a back seat to wages and management practices, were seized on by activists as a critical new front in the Nike campaign. NGOs began coordinating a strategy for pressuring Nike to phase out toxic glues and solvents. For example, NGOs worked with Congressman Bernie Sanders (from Vermont) to draft a letter demanding that Nike resolve these problems. The Sanders letter, which was signed by 53 congresspersons, stated quite forcefully that "As members of the United States

Congress we are deeply disappointed and embarrassed that a company like Nike, headquartered in the United States, could be so directly involved in the ruthless exploitation of hundreds of thousands of desperate Third World workers."

The report was also quoted in the Vietnamese press. At least one Vietnamese paper led its story about Tae Kwang with a reference to the letter signed by the U.S. Congress asserting that the "U.S. footwear giant Nike has been slam-dunked by fifty-three members of the U.S. Congress for exploitive labour practices" (VIR—Nov. 17, 1997). References to the U.S. government's criticisms of Nike were used to make clear that it was not the Vietnamese government or just a few labor activists that were being critical of the company. This focus on U.S. criticisms also helped open room for further government inspections of Nike subcontractor factories. Local regulators felt freer to evaluate environmental and health conditions in Nike factories after these problems became public.

The Ernst & Young report was also distributed widely through the Internet. A year after its release, the Web page providing the Ernst & Young report was still receiving approximately 1,000 hits per week. Nike responded with web pages of its own attempting to counter the critical information by advancing the "facts" about its factory conditions. The Ernst & Young audit and other NGO research has also been used by what Wapner (1996) calls "nonformal channels" of pressure such as Doonesbury cartoons spoofing Nike for its poor practices and a Michael Moore film called *The Big One*, which made Phil Knight—the CEO of Nike—appear both callous and out-of-touch with the conditions in his factories.[5]

A series of investigative reports were also motivated by the Ernst & Young audit. A team from ESPN traveled to Vietnam to analyze the conditions at Tae Kwang Vina (and other shoe factories). ESPN eventually produced a one-hour documentary on sweatshop issues in the shoe industry that included the interview with the worker who was hit by her boss. This documentary, which is probably the most political topic the sports channel has ever covered, was particularly stinging to Nike as it reached the company's primary target audience—young, male sports enthusiasts.

The Ernst & Young report also was used as evidence in a lawsuit filed against Nike in the state of California in April 1998. A law firm that had previously sued "Joe Camel" (the cartoon character that advertises Camel-brand cigarettes) sued Nike under similar allegations of consumer deception and false advertising. On the day the lawsuit was announced, NGOs used the media once again to raise public awareness about issues of workplace conditions at Nike factories.

As Keck and Sikkink point out, "The ability to generate information quickly and accurately, and deploy it effectively, is [an NGOs] most valuable currency; it is also central to their identity" (1998: 10). Through direct experience in this campaign, it is clear that NGOs seek to access accurate and defensible information that supports their position, repackage it for a mass audience, disseminate it through multiple channels, raise it repeatedly, use it to motivate others to take action (such as asking congresspersons to sign the Sanders letter and consumers to buy differently), and incorporate it into other strategies (such as the lawsuit). These groups are now able to gather information and transfer it incredibly quickly from Vietnam to the United States to the rest of the world. Even though their information gathering is somewhat haphazard and unsystematic, it has been effective in publicizing the most egregious examples of workplace and environmental abuses at Nike factories.

6.4 The Media

The media plays a critical role in non-governmental strategies. Even in Vietnam, where the press is controlled by the government, newspapers contain a surprising number of stories on pollution, labor, and development issues. If their headlines are any indication, Vietnamese papers are not bashful about raising politically sensitive environmental issues. Just to give a flavor, newspaper reports regarding the case study firms included headlines such as: "Pollution: The Seamy Underbelly of Development in Hanoi;" "Factories Generate Serious Pollution: To Shut Down or Not to Shut Down?"; "Ministry Alarmed over Surging Chemical Pollution"; "When Will Thanh Mieu Ward Escape from a Polluted Condition?"; and, "Dona Bochang Factory Continues to Generate Pollution."[6] Headlines regarding the Nike case were even more direct: "Nike accused of Vietnam worker abuses"; "Foreign bosses blasted—labor force abused as firms seek short cut to quick buck"; "Nike fails to kick abuse allegations"; "Taiwanese woman jailed for maltreating workers"; and, "Nike to axe sweatshop contractors."

It appears that reporters have more freedom to write about environmental conflicts (which often represent deeper political or economic conflicts) than other political issues (such as peasant protests or government corruption). It is not clear why newspapers can openly report about corporate pollution and mistreatment of workers (and underlying state inaction or complicity) while other press reporting is sternly controlled.[7] It may again return to the issue of state legitimacy and the state's proclaimed role in protecting society from the problems of capitalist development. For whatever reasons, though, the press continues to report on pollution and to play a critical role in pollution debates. The international media is even freer to report on controversial issues and topics that are potentially embarrassing to the government.

Media reports about pollution problems are often the only source of information that communities can access. Very little environmental information is made public by the government. State-of-the-environment reports remain confidential, as are inspection reports and monitoring data. The government's reluctance to release environmental information is based allegedly on concerns that environmental reports will falsely state a problem or accuse a firm of pollution that cannot be supported, and that this information will lead to conflicts between communities and firms. It also appears there is political pressure against public reporting from some state enterprises and government agencies.

Nonetheless, because so little official environmental information is made public, media and NGO reports are particularly important for pollution campaigns. First and foremost, these reports pressure state agencies to respond in some way to specific environmental concerns. Members of virtually every community I interviewed talked about the repeated complaint letters they had written that elicited no response. However, after a newspaper article or television report came out, these very same state agencies began to respond to their complaints. Even the highest levels of the government such as the Prime Minister and National Assembly representatives feel pressure to resolve local environmental issues when they are reported in a local or national newspaper.

Media and NGO reports often directly or indirectly call into question a state agency's effectiveness and commitment. Newspapers such as *Lao Dong* (Labor) and *Thanh Nien* (Youth) are particularly effective at challenging agency actions and performance on issues that affect their constituencies. Investigative stories that chronicle agency inaction, or that tell the story of a community that has repeatedly complained to no avail, embarrass government officials. An article regarding the Dona Bochang case, for instance, sent a direct message to local government officials with the long, but targeted headline: "Hope That Concerned Agencies Will Make Efforts to Solve the Environmental Pollution at the Dona Bochang Factory Area."

Media and NGO reports can also redirect pressures onto other state agencies, creating Keck and Sikkink's "boomerang" effect (1998), or what might be thought of as "venue shifting." NGOs in Vietnam often avoid directly challenging state agencies, or trying to directly alter government policies. Instead they seek out other targets in the state that can put pressure on the key agency or policy. For example, community members around Ba Nhat sought to pressure their National Assembly representative after they realized they were getting little response from Hanoi city officials. Through the press, they were able to raise the local pollution issues with higher agencies and politicians, and to expand responsibility to other actors.

When other agencies, or different levels of the government are brought into the debate, it is sometimes possible to alter a local agency's calculation of costs and benefits of action. For instance in the case of Tan Mai paper, reports of high levels of pollution from the factory had reached the Prime Minister and forced a response (even if it ultimately failed) at the local level. Agency officials responsible for dealing with Tan Mai's pollution were nervous they would get a call from the Prime Minister's office or even worse a visit to show his responsiveness to public concerns (as he has done with other factories).

The media and NGOs also apply pressure directly on firms. Most companies seek to foster a positive public image, or at least to avoid negative publicity. Newspaper and television reports can easily stain a company's reputation, particularly for name-brand manufacturers, which is much harder to reclaim than to protect. Stories pointing out company disregard for community members, workers, or the environment are embarrassing to both state-enterprises and multinationals.

Finally, media and NGO reports also serve to transform debates about environmental issues. For instance, the media can reframe a story about dust pollution into a health issue, transforming the problem from simply an aesthetic issue (focusing on smells and appearance) into an issue of toxics, health impacts, and long-term ecological implications. Knowledgeable reporters are critical to this reframing.

The media played an important role in several of the cases. In the Dona Bochang case, the media buoyed the claims of community members, and in the process put pressure on both the Dong Nai People's Committee and the factory. A number of newspaper reports and at least one television segment served to legitimate community complaints. Video images of pollution bellowing out of the factory's smokestacks were shown to reporters and government officials, and photographs taken by community members were provided to newspapers. One neighbor explained the reason their complaints had been successful was that while the factory was not afraid of the government (who is a partner in the factory), "Dona Bochang is afraid of publicity." This publicity of course hasn't occurred naturally. Community members have pushed, cajoled, and begged the media to report on the factory's pollution and the state's failure to regulate it. As one community member explained (Mr. Doanh—6/7/97), "I have been complaining for six years so the editor of the newspaper knows me well."

In the Tan Mai case, the media has also reported on the factory's pollution, but much less frequently and without assigning blame to either the factory managers or the local regulators. These newspaper reports have nonetheless had some impact. As one community member explained (Mr. Dung—6/2/97), "the *Phuong* had a meeting with the factory about the pollution which was caused by a newspaper report." And further, "the decision [to build a waste treatment plant] was caused by people who complained to the newspaper and to the *Phuong*." As a factory manager asserted (Mr. Binh—3/12/97), "If the community complains, that is a pressure on the government to investigate us. Mr. Vo Van Kiet [the Prime Minister at the time] always reads the newspaper. So when people write a letter it is dangerous for us." Although these media stories have still not motivated Tan Mai to reduce its pollution, they did pressure the local People's Committee to publicly demand that Tan Mai commit to building a waste treatment system.

Community members around Ba Nhat actually wrote their own articles about the pollution problems and paid to have them run in local newspapers and magazines. The community also worked to convince reporters to come out to write stories. Community members believe (Mr. Tien— 5/28/97) that "the media has been an important factor, showing everyone in Hanoi and the country about the problem. This has helped to accelerate the process. . . . There has been a positive effect from the TV and newspapers."

6.5 Motivating Changes at Nike and Tae Kwang

Advocacy campaigns against firms such as Nike have achieved some important changes in Vietnam. NGO representatives have recently gotten to sit down at the table with Nike to discuss labor and environmental issues. While only a year before, Nike would not admit problems existed in its factories, because of the multifaceted campaign that continued to tarnish its image, Nike agreed to discuss problems with advocacy organizations and to take steps to prove it had resolved them.

In what would appear to be a victory for transnational advocates in May 1998, Nike announced a six-point plan for improving conditions at all of its factories, including a guarantee that all factories would meet U.S. Occupational Safety and Health Administration (OSHA) standards. If Nike fulfils these promises, the transnational coalition of NGOs that has worked to pressure Nike for the last few years can honestly claim credit for having improved the lives of workers in places like Dong Nai, which few of them are likely to ever even visit.

The question, of course, remains whether Nike will implement the promised changes. NGOs have thus been pushing Nike to accept a system of independent monitoring that would involve locally based NGOs, labor groups, and religious organizations to evaluate its factory practices. Based on NGO pressures for independent monitoring, and continued questioning from reporters, Nike agreed in late 1998 to allow me back into the Tae Kwang factory to evaluate conditions.

In the eighteen months after the release of the Ernst & Young report and the New York Times article, Tae Kwang had implemented important changes which appeared to have significantly reduced worker exposures to toxic solvents, adhesives, and other chemicals. The factory managers also seemed to have gained a much better awareness of workplace health and safety issues, were more committed to implementing safety and health systems, and had assigned responsibility for these issues to a top manager.

Significant investments had been made in the transition to primers, adhesives, and cleaning agents with reduced volatile organic compound (VOC) content. These changes had resulted in: near complete elimination of toluene; near complete elimination of acetone; significant reduction of methyl ethyl ketone (MEK); and complete elimination of methylene chloride. A range of new compounds with reduced organic solvent or natural solvent content had replaced the chemicals that were previously causing respiratory problems.

Tae Kwang had focused most of its efforts over the eighteen-month period on the reduction or elimination of worker exposures to airborne chemicals in the workplace. This prioritization made sense from a risk perspective and from the perspective of responding to public critics. Other improvements however, had also been made, including:

- Ventilation systems had been installed.
- Tae Kwang had partially implemented a personal protective equipment program.
- Tae Kwang had installed a new incinerator which was said to be more efficient and less polluting.
- Tae Kwang had expanded its material recycling program, which consists of sales of scrap to off-site companies that either reuse, repackage, or recycle the materials.
- Tae Kwang had improved chemical delivery and handling. The company had made major changes in the chemical mixing area—constructing rooms with hoods to do mixing, and providing workers with respirators and lab coats.

Since this reaudit, activists around the world have continued to pressure Nike to accept independent monitoring of health, safety, and environmental conditions in their footwear and garment factories. Activists have also redoubled their campaigns to pressure Nike to pay a "living wage" to its contract workers.

6.6 Effectiveness of Nonstate Pressures

Much has been written about the role of NGOs in environmental struggles in the developed world (Szasz 1994, Gould et al. 1996, Wapner 1996) and the impact of nonstate actors in domestic and international politics (Risse-

Kappen 1995, Keck and Sikkink 1998). Analysts have heralded these groups as the harbingers of new forms of "world civic politics," (Wapner 1996) and as the very basis of an emerging "global civil society" (Lipschutz 1996). Although this literature does not describe the specific types of organizations or processes one finds in Vietnam, it does present many of the key dynamics at work in environmental debates, highlighting the need for, and responses of, nonstate actors. I thus draw on these literatures to provide a context for examining the actions of domestic and international nonstate actors in Vietnam.

Gould et al. (1996) for instance argue for the importance of extralocal actors in local environmental struggles, painting a rather bleak picture of the potential of communities to win environmental protections on their own. As they explain, "local citizen-worker environmental movements face considerable resistance from economic and political actors in their locality and their region. . . . Powerful multinational firms use their promise of both jobs and taxes to persuade other economic and political individuals and organizations to bend to their wills" (p. 173). This results in a dynamic where "the transnational mobility of capital makes it more and more difficult for local, regional, and national governments to control their own economies and their own natural environments. Greater fluidity is leading to a very real erosion of state power on all levels" (p. 174). From their case studies in the United States, Gould et al. assert that when faced with community complaints or environmental demands, corporations "frequently wait out and wear down their opponents. These battles are clearly wars of attrition, in which citizen-worker groups frequently need extralocal help in order to stay the course" (p. 36). Summing up their analysis of U.S.-based environmental struggles, they assert that "most commonly, local mobilization dissipates after it fails (or, less commonly, succeeds) in its specific local conflict. When it does survive beyond the conclusion of its initial local mission, this is often due largely to the social resources provided by extralocal national and/or regional environmental social movement organizations" (p. 40).

Faced with this dismal record, Gould et al. argue for the importance of coalitions that connect across different levels of political activity, manage the challenges of knowledge production and dissemination, and help sustain local resistance. The authors assert that a more promising model of mobilization involves community embeddedness in larger extralocal networks. They conclude, "mobilization can be more successful if local communities organize extralocally and monitor locally" (p. 196), with significant resources devoted to building "extralocal alliances in order to confront the extralocal control of national and transnational . . . agents" (p. 215). Milofsky (1988: 20) similarly argues that community groups seldom win on their own, and thus the "main task" of local movements is to "locate resources available in the broader society and attract them to the locality."

Other analysts have also pointed to the importance of connections between local and extralocal actors for successful environmental campaigns. Barnet and Cavanagh's (1994) "globalism from below" requires a local constituency with connections to broader resources and movements. Karliner's (1997) "grassroots globalization" involves connected communities across regions and countries that range from individual consumer pressures, to corporate accountability campaigns, to broader efforts to reform state and international institutions through activist networks.

Keck and Sikkink (1998) describe and analyze a specific form of non-state actor—the transnational advocacy network—which serves some of the functions Gould et al. argue are critical. Keck and Sikkink claim that "where the powerful impose forgetfulness, networks can provide alternative channels of communication . . ." and "multiply the voices that are heard in international and domestic policies. These voices argue, persuade, strategize, document, lobby, pressure, and complain" (p. x). For Keck and Sikkink, "the core of network activity is the production, exchange, and strategic use of information," which serves to "reframe international and domestic debates, changing their terms, their sites, and the configuration of participants" (p. x).

Keck and Sikkink are particularly interested in "the ability of nontraditional international actors to mobilize information strategically to help create new issues and categories and to persuade, pressure, and gain leverage over much more powerful organizations and governments. Activists in networks try not only to influence policy outcomes, but to transform the terms and nature of the debate" (p. 2). Keck and Sikkink also raise the interesting concept of a "boomerang" effect in which pressure is redirected from one agency to another in order to find a point of leverage over the state agency that is responsible for a pollution decision. For example, if one state agency blocks the demands of a local group, community members can enlist the support of extralocal, nonstate actors (such as a sympathetic NGO or reporter) to apply pressure on another agency in the state that could then pressure the original agency to take action.

Wapner's (1996) analysis of environmental NGOs is also relevant to the Vietnamese cases as activists there aim not to establish new governmental agencies or to dismantle existing ones, but to enlist state resources and capabilities. As Wapner notes, "The idea is to discover unconventional levers of power and employ nontraditional modes of action that can affect, if only unevenly and imperfectly, the global community. This involves teasing out and utilizing nonformal channels and mechanisms of political engagement or, put differently, manipulating forms of power that are generally considered ineffectual in the larger context of so-called genuine politics" (p. 159). These groups often work through persuasion rather than coercion, although when deemed necessary they also employ forms of public pressure.

Nonstate actors and extralocal pressures played a significant role in four of the cases examined in this book. In the Nike case it was clear that without their presence the company would likely not have improved its performance, and certainly not nearly as quickly. But it is not as clear in the other cases exactly whether nonstate actors "make or break" successful community-driven regulation. The impact and effectiveness of nonstate actors is a function in these cases of how they interact with other actors involved: firms, state agencies, and communities.

The nature of the firm matters in whether non-state actors are influential. Companies must be concerned about their reputations for information and accountability tactics to be influential. As Keck and Sikkink argue, "target actors must be vulnerable either to material incentives or to sanctions from outside actors, or they must be sensitive to pressure because of gaps between stated commitments and practice. Vulnerability arises both from the availability of leverage and the target's sensitivity to leverage; if either is missing, a campaign may fail" (1998: 29). For example, a firm selling high-profile consumer products that depend upon image and reputation is more likely to respond to media or NGO information about poor performance. Firms selling to other firms that are concerned about their reputation or standards, such as purchasers that require

ISO 14000 certification, will also be more sensitive to critiques of environmental performance. Firms that need the support of local officials, or that want to avoid stricter state regulation, may also seek to avoid negative publicity, as will companies that strive to be respected in the marketplace. In these regards, market reputation clearly matters the most to the two foreign case study factories: Dona Bochang and Tae Kwang Vina.

But reputational impacts appear to have different effects on different kinds of companies. Foreign multinationals seem to fear that reports of pollution or workplace environmental problems will lead to losses in sales among concerned consumers, that they may get inspected more often because inspectors now know to monitor them, and that they may face more red tape from the Vietnamese authorities. State enterprises on the other hand do not seem to fear either consumers (as most of their sales are to other state enterprises) or regulators, but rather state decision makers who approve soft loans and bank financing. As in the case of Viet Tri Chemicals the main fear of managers was that reports of pollution problems and adverse impacts on community members was going to make it more difficult for the factory to access loans from the government to upgrade its production.

Vietnamese companies that export their products to the United States, Europe, or Japan may in the future face increasing consumer demands and pressures for altered production practices. International NGO campaigns can impact these firms' decisions. Multinational firms based in developed countries similarly may come under pressure for the practices of their subcontractors in Vietnam. This is happening in Europe regarding garments, in Japan regarding food products, and in the United States regarding labor issues for a range of products.

The position of a state agency responsible for regulating pollution also matters. Accountability politics work particularly well on agencies that have claimed they are protecting community members or the environment, or that have certified industrial activities are environmentally clean (as in the case with Environmental Impact Assessments). Agencies and government authorities (such as the provincial People's Committees) that need to legitimate their power are also susceptible to attacks on their effectiveness. Agencies that appear to have made money off permitting pollution may also be laying themselves open to media and NGO scrutiny.

The characteristics and capacities of the community also influence whether nonstate actors are effective. If community members can take advantage of the resources and benefits provided by extralocal organizations, nonstate actors will play a synergistic role in environmental struggles. If communities are too weak to utilize media or NGO efforts, reports and articles may serve little purpose. Community members that are good "spokespersons" for their cause, providing a compelling voice to outside audiences, will in turn support the work of NGOs and the media. Outside NGOs are always in need of local legitimization. Solid community support can provide this legitimacy.

6.7 Conclusions

The Nike case shows quite clearly that nonstate actors can win campaigns targeting global firms, even when they are located deep inside industrial zones in the middle of Vietnam. In an environment that leaves little space for NGOs or other intermediary organizations, transnational networks still can play an important role. Taking advantage of global linkages, they can exercise leverage on behalf of local demands that trump the power of firms such as Tae Kwang Vina that appear invincible in the local context. However, these organizations rarely win battles on their own. As Gould et al. (1996: 181) argue, a "successful strategy of mobilization must have a local component." The linkages that groups like Vietnam Labor Watch and Global Exchange have built on the ground in countries across Asia are critical to their information gathering, their political legitimacy, and ultimately their campaign effectiveness.

And while the Nike case is encouraging in a number of regards, it also hints at a host of limitations of these types of transnational processes and campaigns. Clearly one of the key points of leverage in the Nike campaign arises from Nike's brand profile and sensitivity to public embarrassment. If Tae Kwang had been producing a generic brand of shoes, it is unlikely that the factory's conditions would have merited even a footnote in the New York Times. Transnational strategies depend on the media and public opinion for much of their impact. It is thus an open question whether these processes can influence companies that are not tied directly to high-profile brands or retailers.

NGO actors and media outlets also have their own interests and goals in these campaigns, which may or may not match with local community and worker interests. For instance, focusing on Nike may be popular this year for NGOs in the United States (because it is a winnable campaign or a hot topic), but may be old news next year. And even if NGOs in the west continue their interest in Vietnamese factory conditions, their funding may shift to a new topic that leaves Nike workers and their health problems behind.

The Nike case shows both the power of transnational advocacy networks and the fragility of their campaigns. It shows that a sort of "Activist-Driven Regulation" can be effective in pressuring firms in Vietnam to change their practices, but the success of these campaigns depends on many factors spread out across several continents, and is hardly something that Vietnamese workers and community members can count on. Although these networks are fragile in a number of regards, they also hold out some potential for responding to current trends in the globalization of production. For if systems can be built in which Vietnamese workers and community members are empowered to pull a proverbial fire alarm, and in which international NGOs and the media can respond effectively to these alarms, these processes may present a hopeful alternative to counting on local state agencies to strictly regulate the industries they so desperately struggle to attract and promote.

A system of international independent monitoring of subcontractors of Nike, Reebok, adidas, and their competitors could establish the kind of broad net that is needed to control, or at least track, the actions of multinational firms around the world. By pressuring firms such as Nike—and a growing pool of electronics, toy, and clothing brands—to pressure their subcontractors, community members, workers, and international activists may also be able to create some room for local regulators such as those in Dong Nai province to do a better job of inspections and enforcement.

Clearly there is enormous potential for these types of information-based strategies for promoting regulation of environmental, health and safety, and other social issues. If international NGOs take their lead from local concerns, and transfer sophisticated strategies for collecting and disseminating information back to local groups, these campaigns can significantly influence industry practices.

7
Community-Driven Regulation: Community Alarms and State Responses

The preceding chapters have presented a range of detailed cases of industrial activities and their environmental impacts, and more to the point, state, NGO, and community responses to improve environmental and health conditions. My purpose here, however, is not to simply present the intricate histories of these cases, or to tell good stories, but rather to draw lessons from the varied cases. This chapter steps back from the details of the cases with the goal of exploring more general explanations and hypotheses about the underlying processes at work in conflicts over pollution and efforts to balance development and the environment, building on the empirical evidence of the cases to advance a simplified model of the processes supporting more effective environmental regulation in Vietnam.

Comparing widely varying cases—from a modern Nike shoe factory to an old state-run chemical plant—can be a precarious task. Each of the cases is unique in its own way. However, these cases do reveal a number of patterns. Within each complex case, similar underlying dynamics emerged which supported effective regulation. Existing theories do not adequately explain these dynamics in a country like Vietnam. It was not simply the strongest agency, strongest community, or the use of market incentives that determined effective responses to pollution problems. In each case, a combination of factors was at work. This chapter attempts to generalize these dynamics through a simplified model.

7.1 Community-Driven Processes

Processes currently exist in Vietnam which can motivate state and firm actions to reduce pollution. In the cases analyzed here, community actions played a key role in driving these processes—pressuring the state to take action against polluting firms. Of course not all communities were able to mobilize effectively and not all state agencies were responsive to community demands. Nonetheless, some communities were able to mobilize, pressuring firms and the state, and securing important pollution reductions.

From interviews with frontline government regulators it is clear that the vast majority of regulatory actions in Vietnam have occurred only after community complaints. Staff at the three key environmental agencies—the Departments of Science, Technology, and Environment (DOSTE) in Hanoi, Dong Nai, and Phu Tho—admitted that virtually all inspections to date had been driven by community complaints.[1] Representatives of the National Environment Agency (NEA) similarly acknowledged that the inspections conducted up through the time of this research had been instigated after community complaints.

In a certain sense, community pressures have become a sort of "backbone" for state regulatory action. This community backbone function is certainly not unique to Vietnam, nor entirely surprising. Vietnam developed its environmental laws and regulations extremely quickly in 1993 and 1994 with the assistance (and prodding) of international aid agencies and NGOs. The Vietnamese regulatory system has thus grown the outlines, or skeleton of a governance system, with little of the muscle needed to make it stand up. Weaknesses in environmental agencies and the countervailing strength of community demands has led to a situation of essentially community-driven inspections and enforcement.

Community pressures play a critical role in supporting the implementation of environmental laws in Vietnam. Communities pressure the state to enforce existing laws on the books. Demands for effective enforcement of these regulations help build the legitimacy and ultimately the capacity of fledgling state environmental agencies. Community interactions with the state help support a kind of "coproduction" of regulation.

Community mobilizations influence both the occurrence of state regulatory actions and the effectiveness of these actions. Community pressures (combined with extralocal pressures) have contributed to the generation of new environmental laws in Vietnam and more importantly to the practical implementation of these laws. Analyzing the relations between community actors and state agencies, among state agencies, and between state agencies and firms is critical to assessing the effectiveness of actual regulation.

Community members in general are much more interested in results—that is, pollution reduction—than in inspections, reports, EIAs, or even agreements to build treatment plants. Mobilized communities thus serve as a dispersed team of monitors to follow-up on inspections and promises of improvement. This is particularly important, as monitoring and enforcement are the Achilles' heel of traditional top-down environmental regulation.

Community participation can also help overcome other common limitations of traditional environmental regulation. For example, among the tensions that always exist within the state regarding environmental regulation, the most basic is the conflict between the desire to promote economic growth (either by attracting foreign firms or supporting state-owned enterprises) and the countervailing pressure to regulate the adverse impacts of industry. On a microlevel, there are significant incentives (both direct and indirect) for government inspectors not to enforce environmental regulations. There are also often severe constraints on staff and funds, which make regulation difficult even for the most committed inspectors. Community participation in the regulation process helps to tip the balance of this equation towards enforcement. At a minimum, community actions make it more difficult for firms to bribe local officials or to falsely claim that problems have been solved.

7.2 Regulatory Dynamics

Successful cases of Community-Driven Regulation follow a similar pattern:

1. Communities identify priority environmental problems and instigate action to solve them—usually through complaint letters to a local government agency, letters to the firm, or protests;

2. The state responds by investigating, gathering data, and analyzing the past performance of and existing requirements on the firm;

3. The state may also impose fines or require technical changes inside the factory;

4. The community monitors the state's actions (albeit through unscientific means) and any changes in the performance of the firm;

5. If pollution is not reduced, the community escalates its pressure on the firm and challenges the state to fulfill its legal mandate, often turning to extralocal actors such as the media, NGOs, or higher governmental bodies to support their claims.

This pattern of environmental regulation differs in a number of regards from both traditional command-and-control (CAC) regulation and simple public participation models. Under command-and-control, a "patrol" model[2] is employed in which the state sets environmental standards, establishes inspection systems, patrols for violators, and then enforces its solution (Kraft and Vig 1990, Fiorino 1995, Gottlieb 1995). The command-and-control model assumes that the state is an autonomous actor with the power and will to regulate strictly and to patrol firms effectively. Even when command-and-control works, however—and it often does not in developing countries—it has many limitations, the most important being monitoring and enforcement (Afsah, Laplante, and Wheeler 1996, Desai 1998).

In traditional public participation programs, government agencies allow communities to provide input into environmental issues, but the state sets the agenda for discussion and creates the forums for participation (Fiorino 1995, Canter 1996). Governmental agencies also sometimes use participation as a means to protect firms and the state from outside pressures (Taylor 1995). Public participation processes in the United States and Europe for instance, have been criticized for their limited scope, their late-stage procedures when decisions often have already been made, and the reactive position they force communities into (Chess 1999). Communities are increasingly demanding deeper, earlier, and more substantive forms of participation. The environmental justice movement in the United States for instance has harshly critiqued the current system of public participation and advocated mechanisms for communities to have a real say in decisions (even technical decisions) that affect their lives and the environment (Bullard 1994).

Under Community-Driven Regulation, the community largely determines its own environmental priorities and inspection needs, serves as additional inspectors, monitors progress, and increases the accountability of state–firm negotiations. One weakness in this process, as I have mentioned, is that community members focus on environmental problems that they can immediately see, smell, or feel. Impacts also often have to exceed a threshold level to be noticed (e.g., deaths or illnesses) and by this point, significant damage has already occurred.

Community actions—to identify problems, monitor factory performance, demand results, and verify improvements—can also support state actors who are interested in implementing environmental policies. In the best case, Community-Driven Regulation can result in a "virtuous circle" whereby community actors pressure an environmental agency to take action, which results in pollution reduction, thus bolstering community demands and concerns for a cleaner environment, which in turn leads to further community actions and state responses.

For Community-Driven Regulation to be effective, community, state, and firm actions must come together in ways that are mutually reinforcing. Communities must mobilize effectively, they must have a point of leverage within a state agency, and firms must be responsive to state and community pressures. These characteristics and actions are influenced by history, political opportunities, and existing structures of mobilization, as well as by the relations developed among firms, communities, and the state.

7.3 Community Alarms and Reluctant Firefighters

The process of Community-Driven Regulation represents what might be called a "fire alarm" model of environmental regulation on its surface. Recognizing that the state cannot patrol all firms, and in some areas does not effectively patrol any firms, community members pull alarms to target and motivate state action on specific firms. It is never as simple as pulling an alarm of course. The alarm must get the attention of a state actor with power, and direct that power towards specific actions.[3] In order to trigger this action community members must effectively mobilize, strategically act, and often enlist outside forces to support their demands.

To extend the metaphor, alarms vary in "volume" and "timing." Some alarms are pulled only after environmental damage has been done, such as complaints about crops that have been destroyed by pollution. In these cases the alarm is meant to motivate some form of relief or compensation. Other alarms are more like smoke detectors, alerting authorities to impending problems, and seeking to avert serious pollution incidents.

Responses to alarms vary as well. Some alarms are extremely effective, directing an environmental agency to priority environmental issues. Other alarms, however, are ignored. In essence the alarm volume can go up or down, depending on the ways the alarms are communicated. The media and NGOs can help amplify the alarms, while other interests may help drown out alarms. Communities may also give up pulling alarms if they feel they are not being heard.

In the best case, state environmental agencies act as committed firefighters (rather than corrupt police patrols) working to respond to community alarms. Agency connections to the community and a sense of professional responsibility, drive staff to improve their capabilities and performance. With the right infrastructure—efficient alarms and appropriate firefighting equipment—the environmental agency can become quite effective in controlling pollution. If this system of alarms works, firefighters can graduate over time from simply responding to alarms to working to prevent problems through enforcement of standards, spot checks on facilities, education and training, hazard prevention, and responding to tips on violations before problems occur.

Responding to these public alarms can also help build the capacity and the stature of young environmental agencies. At the same time, public alarms and unannounced inspections make factory managers more accountable to environmental laws. The threat of being targeted for action creates incentives for factory managers to invest in prevention, and to avoid the reputational and financial costs of publicity about pollution incidents.

Communities are transformed in this process as well from victims to monitors. Over time, community members can progress from pulling alarms after a problem has occurred, to pulling alarms that can prevent larger problems. If community members feel they can affect change, and that their alarms are being heard by the state, they will be motivated to monitor even more closely.

7.4 "Demand-and-Control" Regulation

Community-Driven Regulation does not replace or even circumvent traditional command-and-control regulation, but rather represents a supplementary community "demand-and-control" regulation. In Vietnam, as in most developing countries, the state rarely "commands" industry against its will to change environmental practices. CDR thus begins from a different starting point than the model of top-down environmental regulation, and involves different drivers of action.

The processes underlying Community-Driven Regulation support a process that might be thought of as "populist maxi-min regulation" (see Fung and O'Rourke 2000). The process is populist in that there is a key role for ordinary people. It is "maxi-min" (following Rawls 1971) in that maximum attention is often focused on the worst conditions. The process begins when community members mobilize to gather, analyze, and disseminate information on a pollution issue. Community members evaluate and prioritize their concerns, often focusing on issues that have not been dealt with effectively by existing state regulation. Communities then make demands on the government and put direct pressures on firms. Finally, if they are effective, communities follow-up and monitor changes.

This demand-and-control process increases the likelihood that existing command-and-control regulation is implemented, and at the same time advances other mechanisms of environmental regulation. Community-Driven Regulation forces deliberation on environmental issues, bringing pollution problems—and the trade-offs that are inherent in these issues—out into the light of public debate. CDR raises basic questions of fairness—essentially who benefits and who pays the costs of a particular factory and its pollution. And Community-Driven Regulation advances a form of accountability politics, raising the question of why the state is or is not doing its job.

Existing regulatory institutions are enhanced by these public processes. As discussed in chapter 5, community pressures can help build the capacity of state agencies, and support their demands for greater resources, staff, etc. So although community-state interactions may appear confrontational in Community-Driven Regulation, they may lead to benefits for both sets of actors, and ultimately to the coproduction of environmental regulation (Ostrom 1996).

7.5 Actors in the CDR Model

Four sets of actors take center stage in these community-driven processes: (1) community members affected by the pollution from a factory; (2) officials within state agencies responsible for regulating and promoting that factory; (3) extralocal actors such as the media, NGOs, and consumers; and, (4) factory managers. Both the character of each set of actors and their interactions shape the effectiveness of environmental regulation.

7.5.1 Communities

Community characteristics and their relations to firms and the state determine whether communities mobilize, how they mobilize, and whether their actions are effective in pressuring for pollution reduction. Communities are rarely homogenous. Even in the small communities that were the focus of this analysis, people living in the shadow of a factory could be both cohesive groupings of individuals with similar interests, and opposing forces with different levels of power, wealth, and education who battled to advance their own interests.

In evaluating why communities mobilize, and why some mobilizations are effective, it is necessary to analyze both the structure and characteristics of these communities, and how they relate to the actors they are working to influence. Relations with the media and NGOs also influence community actions. As a number of authors have pointed out (see for instance, McAdam, McCarthy, and Zald 1996), political opportunities, mobilizing structures, and framing processes are key to the emergence of mobilizations and the forms they assume. Three factors seem critical to the effectiveness of community pressures in Vietnam: capacity, cohesiveness, and linkages.

Capacity involves very basic knowledge of rights and complaint procedures, as well as more sophisticated strategies for pressuring state agencies and firms. Individuals need to be able to analyze and document pollution impacts (on health, crops, income, etc.), and then mobilize other people in the community to take action. Community members' understanding of ecological and health impacts of pollution influences whether and how they mobilize. Knowledge of legal rights, and even very simply knowing who to complain to, serves as an important basis of community action. Community members need as well to be able to continuously monitor state and firm actions and improvements. These basic capacity issues have been observed by a number of analysts, who have thus hypothesized the influence of general education and income levels on community mobilization around environmental issues (Afsah et al. 1996). Individual capacities obviously also matter in community organization and mobilization. Individual skills, willingness to take risks, charisma, and commitment all play a role in leading community members to act.

Cohesiveness within the community is critical to effective mobilization. Strong social ties within the community help overcome collective action problems and aid in the mobilization of resources for action. Social cohesion, or what Putnam (2000) calls "social capital," Woolcock (1998) calls community "integration," and Portes (1995) calls "strong ties," is at the heart of successful community organization and action. Informal norms and networks, shared interests, shared identities, and the community's history of organization all play a role in community cohesiveness. Vietnam's socialist past, and the system of local People's Committees, has strengthened this social cohesion. It appears that a shared sense of injustice can also support community mobilization and action. Moral outrage is often based on experiencing an affront to agreed upon community values (Szasz 1994), or to an historical sense of rights and traditions. At the same time, some communities may also be highly divided. If a community is too dependent on a firm, such as in a "company town" when a high proportion of the community works in, or benefits from the factory, it can be extremely difficult to mobilize enough pressure to motivate firm changes.

Linkages, meaning external ties to state agencies and extralocal actors (Woolcock 1998), are also critical if community mobilizations are to be effective. Communities need ties to government officials and agencies as points of leverage over environmental decisions, and more generally to advance community interests. Trust and cooperation with local government officials can make or break community attempts to mobilize. Accountability of local government officials, or more cynically, the ability of community members to uncover and block corruption, is critical to influencing state action. The establishment of the National Environment Agency in 1993 and the provincial Departments of Science, Technology, and Environment (DOSTEs) in 1994 created a focus for community efforts to influence state environmental decisions. The DOSTEs in particular not only implement national policy, but have become targets for community demands, and thus are pressured to communicate those demands to higher state decision makers. Close ties between community members and agency staff facilitate a two-way communication process. Linkages to extralocal actors such as the media are also useful for advancing community demands and for reaching higher government authorities.

In short, when communities are endowed with capacity and cohesiveness, as well as certain linkages, the prospects for Community-Driven Regulation are quite promising. Even though most communities are only partially in possession of these characteristics, my cases show a clear relationship between relative endowments along these dimensions and success.

7.5.2 State Environmental Agencies

State environmental agencies are both coherent actors and arenas for competing interests.[4] These agencies have internal political conflicts, must cooperate with other state agencies and higher authorities, and must respond to external pressures. The success of environmental agencies in enforcing laws seems to be tied up in their internal characteristics—such as their own coherence and competencies—and in their connections to society at large, such as their public credibility and external accountability. For state agencies, capacities, linkages, and autonomy appear to be the key characteristics influencing environmental performance.

Capacity for an environmental agency is a fairly straightforward issue. Agencies need to have the basic organizational, fiscal, and human capital to enforce environmental laws. State agencies must be able to evaluate both the complaints of community members and the assertions of factory managers. Staff need to be able to analyze health impacts, crop damage, and the characteristics of the pollution in question (chemical components, volume, duration, etc.), and to conduct "real-time" evaluations. Environmental staff also need the capacity to know what to demand of a firm, what measures to require for controlling or preventing the pollution, and how to evaluate changes in performance. In the best case, one can imagine a sort of Weberian environmental bureaucracy with well trained staff (hired through a meritocratic system), working in cohesive organizations, insulated from the influences of polluters, with unambiguous policies and enforcement procedures, and clear channels of authority. The reality of environmental agencies in Vietnam unfortunately is still far from this ideal.[5] Most environmental agencies are underfunded, understaffed, poorly trained, and have highly ambiguous policies and enforcement procedures.

Linkages are also key to successful regulation. By *linkages*, I mean the social and political connections among state agencies and between state officials and civil society actors that foster effective communication and feedback. Social relationships are at the heart of regulation. State agencies operate at many levels and interact with civil society actors in diverse ways.[6] Because of the important role of linkages, local-level agencies (particularly those that have strong community ties) are more successful in implementing state policies than provincial or national agencies. Of course, social relationships can also be sources of pressure that impede an agency's effectiveness: factory managers working to protect their interests, will also attempt to build and activate ties to the local agency. Linkages must therefore be balanced by autonomy.

Autonomy to enforce environmental regulations, particularly when regulations threaten private or state interests, is exceedingly difficult for young environmental agencies to achieve. By *autonomy* I mean intrastate autonomy as well as autonomy from the private sector. Autonomy requires protection from capture or undue influence from either firm managers or other state officials interested in the promotion or protection of industry (such as the Ministry of Industry or the Ministry of Planning and Investment). Insulation from elite demands requires formal legal procedures and countervailing pressures. Inspectors themselves need to be monitored for accepting gifts and bribes. Divisions between state agencies or between broader groups in society can create the openings needed for this autonomy. Community pressures can also force the state to be more responsive to community demands and thus more autonomous from elite interests.

7.5.3 Extralocal Actors

At the local level, it is common for community members to lose in political battles. In these situations, effective communities often turn to extralocal resources to strengthen their campaigns. The most prominent of these are national and international media and nongovernmental organizations (NGOs). Vietnam is somewhat unique in the kinds of extralocal resources that are available to communities. There are still no truly independent NGOs working on pollution issues. NGOs that do exist are actually government or university-sponsored groups, or international organizations. These organizations are constrained in their activities, and generally do not attempt to influence the Vietnamese government directly, but rather work indirectly to change policies and practices.

Given the limited presence of NGOs and other independent intermediary organizations in Vietnam, one might assume that this category of actor could be dropped from the analysis. The Nike case, however, demonstrates that extralocal NGOs can have a substantial impact, even in Vietnam. And the Ba Nhat case shows that even ostensibly state-controlled NGOs (such as a government-run university NGO) can play important roles in pressuring agencies within the state. The key is leverage and linkages once again. NGOs are effective when they are connected to and have influence over local, national, or global decision makers. Under certain circumstances, NGOs can provide extraordinary leverage on behalf of local struggles.

7.5.4 Firms

Finally, of course, there are the firms themselves. Although in these cases firms are generally sources of pollution rather than leaders of sustainability, it is nonetheless important to conceptualize the characteristics that make firms more or less likely to respond positively when they are put under pressure by communities, state agencies or extralocal actors to reduce their pollution. Analysts have proposed a long list of characteristics which appear to be associated with "greener" firms (see for instance Ashford and Heaton 1983, OECD 1985, OTA 1994, and Lawrence and Morell 1995). My research points to three broad characteristics which enable firms to respond positively to environmental pressures: capacity, linkages, and market positioning.

Capacity for firms refers to the technical and organizational capabilities needed to solve environmental problems. Firms must be able to identify sources of pollution, develop strategies for controlling or preventing the pollution, and overcome internal and external barriers to process changes. Factory managers need to be able to access relevant information (regarding internal practices and external options and alternatives), alter organizational structures to create incentives for pollution reduction (from top managers to engineers to line workers), mobilize capital for investments, implement technical changes, and measure the effectiveness and cost of changes. Internal political structures also influence firm responses to environmental problems. Firms need dedicated environmental staff that have a say in production decisions which affect the environment. Integration of environmental staff with production staff helps alleviate the marginalization of environmental concerns, and supports wider reaching pollution control efforts.

Linkages affect the performance and responsiveness of firms as well. Social ties between a firm and the community can be instrumental in motivating the firm to take action on pollution problems. How a firm is connected to the local community, such as through its workers, consumers, suppliers of inputs, or just as a neighbor, influences how it responds to community complaints about pollution. In cases where community members can influence decisions that affect the firm, such as access to local resources, firms become particularly responsive to local demands. Likewise, if a firm looks to the state for access to capital, import and export permits, or licenses for doing certain activities, the firm may be more inclined to follow state directives on pollution. As in the case of state agencies, however, close ties can cut both ways. When jobs and taxes are at stake, threats of capital flight can protect firms from community and state demands.

Market position affects both the financial and strategic position of firms as they face environmental decisions. If a firm is a market leader, with strong brand recognition, it will likely be more sensitive to public criticism about environmental practices. If a firm sells to an environmentally aware market, such as food products to Japan or textiles to Europe, the company may also come under pressures for improved environmental performance. Conversely, if a firm produces low-margin, intermediary products such as basic chemicals, it will have little market incentive to improve its performance or protect its reputation. A firm's market position also affects state and public expectations for the firm.

7.6 A Schematic of the CDR Model

Figure 7.1 presents a simple schematic of relationships between these actors, showing that a number of routes can be taken to influence firms to reduce pollution. Communities can pressure firms directly, or can work with the media and NGOs to pressure firms. Extralocal actors can magnify or redirect complaints to other state authorities thereby adding new pressures on the state to respond. And the state on its own can pressure firms. The most effective process I examined in Vietnam involved a dynamic in which communities—in coordination with extralocal supporters—pressured the state to pressure a firm. Communities then monitored the implementation of regulations and improvements.

The diagram is drawn to indicate both the direction and the force of interactions in Community-Driven Regulation. The strongest interaction is from the community to the state to the firm. This interaction leads to the most significant impact on environmental performance. This relationship can also be reversed with the firm pressuring the state, which in turn can work to suppress a community's demands. Communities can also directly

Figure 7.1
Community-driven regulation diagram

pressure firms. However, in Vietnam direct community pressures on firms are currently a weaker and less effective link.

These interactions are obviously complex. There are often significant internal conflicts within each set of actors. Relations between actors also vary widely. In the "best" cases, communities are cohesive and connected, state agencies are capable and responsive to community pressure, and firms are accountable to consumers or the state and in a strong market position to respond. With these characteristics, the levers of pressure and response can be fairly straightforward.

However, as we have seen, there are also cases where communities are divided and isolated from state authorities, where state agencies are unresponsive to community needs and corrupted by ties to firms, and where firms are insulated from pressures and motivated only to externalize pollution costs. In these cases demands for environmental protection are often frustrated, forcing communities and NGOs to seek alternative means to pressure the state or a firm.

8

Regulation against the Odds: Conclusions and Policy Implications

Without the support of the public, even the easiest task can't be accomplished. With the support of the public, even the hardest task can be fulfilled.
Communist Party slogan

Regulation of pollution in developing countries appears an almost Herculean task. It is almost taken for granted that governments pass environmental and labor laws and then fail to enforce them, postponing environmental and social protections in the interests of economic development. What is thus surprising is when communities and local governments are actually able to promote more sustainable forms of development against a range of pressures and disincentives.

Vietnam is in many ways a classic example of the challenges of balancing development and the environment. Like virtually every developing country in the world, Vietnam aspires to follow in the footsteps of the "Tiger" economies of Asia. The government is squarely focused on attracting foreign capital, supporting industrial growth, and generally building a modern, industrial nation. However, even with this strong development bias, Vietnam has also been forced to recognize that rapid growth can have serious environmental implications—as seen quite clearly in the neighboring countries of Thailand, Korea, and Taiwan. While representing the potential for successful development, these nations also represent urban environmental problems to be avoided. So as Vietnam begins its development journey with old technologies and new polluting industries, it has the potential to be not just as bad as Bangkok or Taipei, but perhaps even worse.

The recent inflow of foreign direct investment into Vietnam continues to drive a process of urbanization and industrialization. Rapid population growth in cities overwhelms existing infrastructure and exacerbates environmental problems. Traffic congestion (and its associated air pollution), residential overcrowding, unplanned land uses (including the siting of highly polluting factories next to residential areas), uncollected municipal solid waste, and polluted rivers and lakes, are the most visible signs of infrastructural inadequacy and regulatory failure.

Vietnam is poised to repeat the same unsustainable development patterns as its Asian neighbors. However, as a latecomer to development, Vietnam also has the potential to learn from these problems. The government could apply international best practice in policy and planning to prevent Ho Chi Minh City from becoming another Bangkok, implement policies and programs to prevent pollution before it is created, and take measures to control development through better planning, zoning, etc.

At first glance, however, the prospect of enlightened government leadership seems unlikely. The persistence of single-party Communist rule raises the specter of a continued focus on the expansion of production and a disregard for the environmental impacts of industry (as was seen in Eastern Europe). A lack of funds, trained personnel, and political influence severely constrain the effectiveness of environmental agencies. State regulators on their own simply have not been able to control the adverse impacts of industrialization and urbanization.

What about other actors that might contribute to sustainable development: communities, NGOs, or social movements? Again, the prospects seem limited. There are no truly independent Vietnamese NGOs, and the NGOs that do exist are severely limited in the roles they play, appearing much weaker vis-à-vis the state than NGOs in other countries in Asia. Communities cannot count on competitive electoral politics to give them leverage in demanding action from the state. Essentially all protests in Vietnam remain illegal.

Vietnam clearly offers a difficult test for policies or programs to promote more effective environmental regulation and sustainable development.

Given this context, successes of any kind are surprising. And these successes offer potential sources of important learning. The fact that three of the six cases studied in this analysis (Dona Bochang textiles, Viet Tri chemicals, and Ba Nhat chemicals) show substantial changes, and two others (Lam Thao and Tae Kwang) show partial improvements, should be read as evidence that processes do exist in an extremely difficult context for motivating improved environmental regulation. At the most basic level, these cases show that under certain circumstances, community pressures can motivate state agencies to enforce environmental laws. Even in a country without formal channels of public participation, community members can mobilize and demand improved environmental protections. These cases also indicate that community actions may promote wider reforms that actually build the strength and legitimacy of state environmental agencies.

8.1   Existing Community-Driven Regulation

The basic processes of Community-Driven Regulation involve community actions to pressure state agencies to enforce environmental laws. The vast majority of regulatory actions in Vietnam today occur only after these community pressures have been applied. Staffing weaknesses in environmental agencies, inadequate inspection systems, and development imperatives—conditions common to many developing countries—have all limited the effectiveness of traditional state-centric, command-and-control regulation, and have supported the emergence of Community-Driven Regulation.

Community actions influence both the occurrence of state regulation, and the effectiveness of the regulation. Community pressures (combined with extralocal forces) have contributed to the generation of new environmental laws in Vietnam, and to the actual implementation of these laws. Community-driven processes can also help overcome common impediments to traditional environmental regulation, tipping the balance in political battles towards enforcement. This is of course not completely new. Similar processes of local protest drove early regulatory initiatives in the United States and Europe.

However, for Community-Driven Regulation to be effective, community, state, and firm actions must converge in a synergistic manner. The six case studies offer rich detail on how these processes play out within conflicts over pollution in a developing country. Under a comparative lens, the cases also provide insight into sources of variation among communities, state agencies, and firms. Clearly there is a range of characteristics and processes that can either support or impede effective regulation. When a community has strong internal social ties and external political linkages, it has more potential to advance environmental demands. When state agencies can act autonomously and are responsive to community needs, they are more likely to enforce environmental laws. And when firms are both vulnerable to external pressures and in a strong market position, they are more inclined to take measures to reduce their pollution.

One surprising finding of the research was just how vibrant communities in Vietnam were as political actors. These cases challenge the stereotypical image of state-socialist societies in which civil society is crushed and moribund under the overwhelming weight of the state. Communities in Vietnam have a dialogic relationship with the state—particularly around environmental disputes. On the whole, Vietnamese communities appear just as mobilized and combative as their counterparts in neighboring democratic countries. However, as I have mentioned, civil society actions in Vietnam do not take the form of large-scale social movements or political parties. Vietnamese communities mobilize and take actions, but at a scale below the radar of most social movement analyses.

The research also shows that community actions are a necessary but not sufficient condition for pollution reduction. In at least two of the cases (and at different times within others), mobilized communities failed to motivate firms to reduce their pollution. Again, it appears to be the combination and interaction of community, state, and NGO dynamics that leads to effective regulation.

In the cases examined here, however, community actions are often the critical motivating force for the state to monitor firm performance, enforce existing laws, and find new regulatory strategies to achieve environmental goals. Community pressures around individual pollution problems expanded dialogue between communities and the state, raising social concerns about the costs of development and creating new pressures on firms to prove they are providing more benefits than costs to a community. Community pressures helped overcome agency resistance to implement laws that impact other state actors (such as the Ministry of Industry). Community pressure also motivated inspectors to simply do their jobs.

At the same time, state actions also helped support community mobilization. The passage of the Law on Environmental Protection and Decree

26 on environmental fines legitimated community complaints regarding pollution and demands for compensation, even while state agencies were unable to enforce the details of these laws. The creation of the National Environment Agency and the provincial Department of Science, Technology, and Environment (DOSTEs) provided a target for community complaints, even though these agencies initially could not carry out their own mandates.

As the Dona Bochang case shows, community-driven processes can create a kind of "virtuous circle" of environmental regulation. Community demands on the state force environmental agencies to improve their inspection and enforcement capacities (in order to retain legitimacy). As inspections and enforcement increase, communities are buoyed by successes to make greater demands on the state. In three of the cases, state agencies played pivotal roles in supporting and legitimating community demands for pollution reduction. In each of these cases, it was clear that successful community action required identifying and focusing pressure on the right actors within the state.

As Evans (1996) has argued, "linking mobilized citizens to public agencies can enhance the efficacy of government," resulting in a kind of "synergy" when "civic engagement strengthens state institutions and effective state institutions create an environment in which civic engagement is more likely to thrive." The outlines of Community-Driven Regulation in Vietnam display this dynamic of mutually reinforcing interactions between organized communities and state environmental agencies. The characteristics of the individual actors in these processes are, of course, critical.

8.1.1 Communities

While there is no single type of community that succeeds at motivating environmental regulation, a number of features of effective communities emerged from the research, including: cohesiveness within the community (and its resulting ability to organize and mobilize); capacity, in the form of strong leadership and problem-solving skills; and linkages, particularly connections to local government authorities or NGOs and the media. Community members' understandings of ecological and health issues also influence whether and how they mobilize. Essentially, successful cases involved communities with strong internal social ties and strong external political ties that forcefully pressured a state agency over extended periods to take action.

8.1.2 State Agencies

The effectiveness of state environmental agencies in enforcing laws was a function of their internal characteristics—such as their own coherence and competencies—and their connections to society at large. When these agencies feel pressure to act, they can work to deliberate and negotiate over environmental enforcement. Media reports and public letter writing campaigns combine pressure from below, with pressure from the hierarchy above, to squeeze environmental agencies into action. For this to work, however, state agencies need a certain level of autonomy from industrial actors, otherwise, economic interests take precedence over environmental concerns. State agencies also need a certain level of "embeddedness" with the communities affected by the pollution, creating a connection for communities to communicate their complaints and demands.

8.1.3 Nonstate, Extralocal Actors

Nonstate and extralocal actors can also influence environmental regulation, serving as a resource for organized communities. The most common of these are the media and environmental non-governmental organizations. Despite a generally unsupportive environment for NGOs, the Nike case demonstrates that extralocal NGOs can have significant impact. The key to their impact is connectedness, or linkages, once again. NGOs can be embedded not just in a set of provincial or national connections, but in global networks which can provide extraordinary leverage on behalf of local interests. To be most effective though, NGOs need credibility and the ability to get information into the public sphere.

8.1.4 Firms

Ultimately it is firms themselves that make decisions about pollution reductions. Having financial and technological capacity to reduce pollution creates the possibility for firms to invest in technological changes. However, motivation is often more important than simple capacity. Linkages, particularly in the form of social ties between a firm and the community, as workers, customers, and neighbors, can be instrumental in motivating a firm to take action on pollution problems. In cases in which community members can influence decisions that affect a firm, companies become particularly responsive to local demands. Linkages to the state (through ownership arrangements), on the other hand, can increase a firm's ability to resist community pressure. Finally, a firm's market position and brand reputation significantly influence how responsive it is to consumer and public pressures.

8.2 Theoretical Implications

From the case studies, it is clear that environmental regulation is not simply a function of state capacity, or of better functioning markets, or even of the existence of "social capital" in a community. Rather, interactions between state agencies, community members, and firms are critical to the implementation of environmental regulation. It is this action between "zones of politics" as Andrew Szasz (1994) calls them, which most significantly influence how environmental regulations play out on the ground. Understanding these processes requires research that examines and explains interactions across traditional fields of study.

Complex social and political processes underlie state and firm decisions to control pollution. State agencies often face conflicts (within and between agencies) around regulatory enforcement. Firms vary in their decision-making processes and motivations. Communities mobilize for different reasons and in different ways. And extralocal actors can play a wide variety of roles in environmental debates. Examining the interactions between these actors, and in particular the processes which motivate the state to take action to regulate firms is central to understanding industrial environmental challenges in developing countries.

It may seem obvious to argue that environmental regulation is a complex process, involving multiple actors with varied interests and capacities. However, too little theory or empirical research in developing countries has focused on these dynamics between actors in the regulatory process. Environmental regulation is tied up in much more than just state environmental policies. Competing interests within the state, between state agencies, and then with external actors are critical to understanding how regulation actually is implemented. The Community-Driven Regulation framework is one attempt to make account of these complex dynamics influencing regulation in a developing country.

In analyzing successes and failures of regulation, I have sought to understand why the state intervenes at all around pollution issues, when it intervenes, and under what conditions this regulation is effective. My research has also sought to address broader questions about state regulatory processes and how environmental concerns may be advanced during development and to look beyond traditional regulatory strategies to analyze innovations from different sectors and to learn from successful processes where they exist. By trying to understand what currently works, and why, it may be possible to build on these successes and design future policies to more effectively promote environmental protection.

The Community-Driven Regulation framework engages and hopefully advances existing theories on environment–development challenges, by synthesizing and building on theories regarding the interactions between state agencies, communities, and firms around pollution issues. Community-Driven Regulation is helpful in explaining when the state can be pressured to intervene to protect the environment, and when and under what conditions communities and workers can mobilize effectively to oppose degrading development.

Community-Driven Regulation also sheds light on how transnational advocacy networks might play a more systematic role in improving industrial practices. The Nike case hints at how a transnational coalition connecting workers in Vietnam, to activists in Washington, DC, to consumers and concerned citizens around the world can advance strategies to "monitor locally and mobilize extralocally" (Gould, Schaiberg, and Weinberg 1996). Many questions of course remain: on how these networks actually work, how democratic or participatory they really are, and how they can be most effective in changing global business practices.

In a number of regards, the CDR model challenges traditional notions of state-centric environmental regulation. The empirical cases and the conclusions drawn from them engage debates about altering the role of the state to more effectively foster and facilitate effective environmental protection, and in particular to strengthen the role of public participation in environmental protection. These cases show that one key to advancing environmental protection (particularly in countries with weak state agencies and strong development agendas) often lies in dynamics outside the state, but that ultimately must influence the state. Community dynamics can help to motivate, supplement, and innovate regulatory practice.

8.3 Relevance to Other Countries

The lessons of Community-Driven Regulation are relevant beyond Vietnam's borders. The Nike case is the most obvious example. Vietnam is one among many countries working to attract multinational firms such as Nike, and at the same time to regulate them. The Nike case shows an interesting process—clearly still emerging—for regulating a global firm. But the other cases in Vietnam also speak to these broader issues. Each of the firms studied plays an important role in Vietnam's efforts to industrialize as the country integrates into the world economy. The challenges Vietnam faces in balancing development and environmental concerns are similar in many ways to those faced by its contemporaries in Asia, the Americas, and Africa.

And these are not just challenges for developing countries. Governments in both developing and developed countries increasingly face barriers to enforcement of even the most minimal environmental regulations. Pressures of globalization and fears of capital flight force governments to walk a fine line between economic development and damage control on environmental degradation—degradation that may, in fact, be undermining long-term development potentials.

Even advanced industrialized countries such as the United States face these challenges. In 2001, the United States entered an economic downturn. Political leaders responded immediately with promises to protect the economy and to reduce overregulation. As President Bush explained after announcing a series of measures to roll back environmental regulations on industry, "We will not do anything that harms our economy. Because, first things first, are the people who live in America. That's my priority. And I'm worried about the economy" (Gerstenzang 2001). Trade-offs between economic development and the environment are certainly not diminishing in national politics. And current state-centric command-and-control regulation continues to come under attack from industry.

Community-Driven Regulation offers one strategy for addressing this impasse. CDR points the way towards new systems of community participation in monitoring, enforcement, and decision making on enviromental and health problems, where communities, including workers, neighbors, and concerned stakeholders, participate more fully in regulatory decisions. These processes can help move politics beyond current environment versus development debates, to local solutions to difficult problems.

A growing movement around the world—among both governments and civil society organizations—seeks to promote deeper public involvement in environmental regulation. Take, for instance, the case of "bucket brigades" in the United States. This simple tool—literally a five gallon bucket with a pump and sealed sterile bag inside—is being used by predominantly poor and people of color communities living near polluting industries to sample air quality and industrial emissions when accidental releases, leaks, fires, or explosions occur. These community samples are then sent to laboratories for analysis. For the first time, community members have access to near-real-time data on the chemicals companies have been releasing into their neighborhoods. This simple process has significantly changed the terms of public debates about pollution and regulation in the communities in which they have been deployed. Bucket brigades have motivated local regulatory agencies to do their own sampling and monitoring. And they have gotten firms to come to the table to negotiate with communities over pollution control and prevention plans (O'Rourke and Macey 2003).

The "environmental justice" movement in the United States has also led a number of innovative initiatives that have placed community-involvement at the center of regulatory strategies. Community groups have analyzed past regulatory practices (and the unequal enforcement of environmental laws) and pressured state and national officials to increase enforcement in poor communities. They have used existing environmental data to pressure both government and firms directly, employing advanced information systems such as the Toxics Release Inventory (TRI) and Geographic Information Systems (GIS) to analyze pollution data and educate community members about environmental conditions (Fung and O'Rourke 2000). Environmental groups have then taken raw data and repackaged it for public consumption, creating "Dirty Dozen" lists of the worst polluters in a county or state, benchmarking the best and worst firms, and printing maps of pollution hot-spots and cumulative exposures of different communities.

These efforts have made clear that "compliance" with specific standards or regulations is not the issue. Communities are concerned about their overall well-being and risks to the environment and their children's health. They have thus used community-driven processes to pressure for new laws and new investments by industry to prevent pollution and to reduce toxics used in production. Trends in community-driven processes are similarly interesting in developing countries.

The World Bank has conducted a number of studies of pollution control in China, Indonesia, Mexico, and other developing countries, and found results that corroborate the findings of Community-Driven Regulation in Vietnam. Pargal, Hettige, Singh, and Wheeler (1997) show that community-based "informal regulation" takes many forms, including demands for compensation, social ostracism of firm employees, threats of protests and physical violence, boycotts, and monitoring and publicizing a firm's pollution. Around the world, communities now often report pollution incidents to regulators and pressure for stricter government monitoring and enforcement.

In China, for example, community awareness and protests regarding pollution have increased significantly during the 1990s. Community members now file literally hundreds of thousands of complaint letters to government officials each year. Informal community pressures were found by World Bank researchers to increase the frequency of government inspections (Dasgupta, Laplante, Maminga, and Wang 2000), to increase the effective rate of pollution charges (Wang and Wheeler 1999), to be "at least as strong as formal regulation" in influencing firms to reduce pollution, and most importantly, to lead to pollution reductions throughout the country (Dasgupta and Wheeler 1996, Wang 2000).

8.4 Limitations of Community-Driven Regulation

As noted, three of the cases from Vietnam show that Community-Driven Regulation can work. However, each of the cases also illustrates a variety of factors that can stand in the way of effective environmental regulation. Divisions within a community, poor organizing, and the inability to find leverage over a recalcitrant state agency, undercut possibilities for achieving environmental protection. Clearly communities can be arenas of conflict as well as sites of mobilization. Savvy firms and government agencies can capitalize on these divisions.

Even when communities organize successfully, their capacity can limit what they demand and achieve. With little data or training, community members often end up only complaining about pollution problems that they can see, smell, or feel. This results in a focus on localized, short-term, acute impacts of pollution—which nonetheless accounts for a significant percentage of industrial pollution in Vietnam at present. However, this focus severely limits the range of environmental issues that become priorities for state action. With no knowledge of technical alternatives, communities tend also to push for pollution control rather than prevention, simply because their main goal is to stop local emissions.

Despite positive examples, state agencies also regularly attempt to block community action. Protests in Vietnam are often suppressed. The range of community action is thus circumscribed by state decrees. Vague legal rights create barriers for communities to complain, seek compensation, and demand action against a firm. State agencies also sometimes intervene on behalf of a factory, particularly when the factory in question is a state-owned enterprise or joint venture.

Another potential problem with Community-Driven Regulation is that stronger communities may motivate factories to move to areas with weaker, less organized communities. Or they may scare off dirty factories from siting in their area, again shifting the burden of the most polluting facilities to poorly organized communities. This has clear equity implications if community-driven processes actually increase pollution levels in certain areas. It would certainly be a Pyrrhic victory if community mobilizations served to create a few clean factories, while other communities remained polluted.

The limits of community capacity and the potential inequities of a system driven purely by community pressures underscores the importance of strengthening the capacity and roles of state programs, laws, and national standards. At present, environmental agencies at all levels in Vietnam are weak. Strengthening basic environmental procedures at the national level, such as ambient environmental monitoring, collection of industrial emissions data, and state-sponsored research on environmental priorities, thus remains extremely important. More fundamental however, is to strengthen the political position of environmental agencies within the state. One of the optimistic outcomes of community-driven processes is that commu-

nity actions can actually help state environmental agencies build their capacities and political strength.

8.5 Supporting Community-Driven Regulation through Policy Initiatives

Although only in an embryonic phase, Community-Driven Regulation already shows potential to significantly improve environmental regulation
in Vietnam. Existing community driven processes are helping to overcome
impediments to environmental regulation, to set environmental priorities,
increase monitoring, and gradually build state capacity. However, these
dynamics have not, to date, been institutionalized in Vietnamese laws or
regulations. Policies and programs that formalize mechanisms of community input, create greater legitimacy for community demands, educate citizens about their rights, and support local monitoring efforts, would move
Community-Driven Regulation forward. These policies would also help
address the larger challenge of balancing industrial development with
environmental concerns. In the best case, this would involve developing
programs that would be triply endorsed by communities, firms, and state
actors. "Alarm" systems could be improved, as could state capacities for
responding, and incentives could be developed to support factory improvements in environmental performance.

Many would argue that a strong central state is a critical first step towards environmental protections. However, in a context with weak state
enforcement and relatively strong community action—which exists not
only Vietnam, but in many developing countries—Community-Driven
Regulation makes sense as a strategy for strengthening environmental protections. It is clearly not possible to create a U.S.-style Environmental Protection Agency in Vietnam overnight. Nor, would I argue, is it advisable.
It makes more sense for countries to build on existing processes and endowments that are proven to support environmental regulation. Existing
resources and community energies should serve as a basis of initiatives for
environmental protection.

It is also logical to build on the varied capacities of different actors
in pollution debates. State agencies in Vietnam have many weaknesses,
but they can play important roles—such as gathering information and

facilitating community-firm dialogues. Communities can also play pivotal roles—such as monitoring firm performance and environmental and health impacts. By combining the functions and strengths of different actors, it may be possible to more effectively "coproduce" environmental protections.

Intentional policies are needed to strengthen community participation and to increase state–society synergy to improve environmental regulation. CDR must progress from community complaints about existing pollution, to community participation in the decisions which ultimately influence environmental quality. Policies and programs need to be advanced that bring all four sets of actors—government agencies, community members, firms, and nonstate actors—into regulatory processes.

In the long-term, one would hope that policies and programs could do more than just support pollution reductions at individual factories. Policies should seek to promote broader efforts to balance economic development with environmental protection. This lofty goal, I believe, should focus on developing mutually reinforcing programs that help build state agency capacity, public awareness and involvement, and create incentives for firms to innovate and improve environmental performance. These goals would be supported by strengthening formal rights to public participation, providing supportive mechanisms for public participation in monitoring, evaluation, and enforcement, increasing access to environmental information, increasing transparency in regulation and enforcement, building state regulatory capacities, and learning from innovations on the ground.

As mentioned though, it is still critical to protect weaker and overburdened communities. Laws can be developed that protect against environmental "injustices" of disproportionate environmental burdens on certain communities. Governments can also improve their abilities to evaluate cumulative exposures of communities to pollutants and to protect against toxic hot-spots developing in industrial areas.

And there is a need to advance broader industrial policies to control future development patterns of industry. Policies could be used to move firms into planned industrial zones where they are easier and less expensive to regulate. Policies could then be developed to create incentives for combined treatment and control technologies, and to motivate innovations in production technologies and management strategies.

Obviously, this is no easy task. But with creative policies and planning, I believe it is possible to build on existing dynamics towards more effective systems of environmental protection.

International donors and multilateral banks can play an important role in supporting Community-Driven Regulation. Overseas Development Assistance (ODA) projects often have the leverage to motivate greater public access to information and to strengthen public awareness of the trade-offs between economic development and environmental protection. Donors should use this leverage to strengthen public roles in environmental regulation. For example, multilateral development banks could require that the Environmental Impact Assessments (EIAs) of projects they fund be translated into Vietnamese and made public, and more importantly, that communities be allowed to participate meaningfully in EIA decision making.

Aid agencies can also help support local experimentation and innovation on environmental regulation. No single agency or donor has the one right answer for advancing sustainable development. There is an important role for ODA to play in funding projects that nurture new initiatives and grassroots driven processes. There is also a need for developing procedures whereby projects can flexibly be amended as implementers hit roadblocks or find new strategies. Mechanisms for real reflection and innovative responses, a kind of adaptive management for regulatory processes should be encouraged.

Finally, there is a need to build checks and balances into all levels of regulation in order to block corruption and economic biases. Even the best laid plans for community-driven programs can be blocked by petty corruption. Public transparency is a key to reducing this bias. Ultimately, Community-Driven Regulation is about public participation and oversight in state regulation, which can lead to both greater accountability and effectiveness of state actions.

8.6 Detailed Policy Recommendations

There is limited direct experience with CDR-like processes around the world, but I believe there is significant potential for advancing the underlying mechanisms of community-driven processes through intentional policy. Through work on several United Nations development projects (two sponsored by the United Nations Industrial Development Organization and one sponsored by the United Nations Development Program and the World Conservation Union) and a recent project with the National Environment Agency (sponsored by the World Bank), I have worked on development projects that build on existing community-driven processes. Based on these experiences, I believe that targeted policies and programs can support CDR's underlying dynamics.

The key policies that can support community-driven processes and broader regulatory reforms, include:

Mechanisms for Community Input: Strengthening the role of the public in environmental protection by formalizing mechanisms of community input into environmental decision-making and creating greater legitimacy for community complaints. Programs should focus on educating citizens about their rights and creating procedures for community members to play a constructive role in local environmental protection.

Community Monitoring: Community and worker monitoring programs should be recognized and supported through technical assistance that helps community members document environmental and health problems they experience. A program of community bucket brigades, which lets community members collect samples when they believe conditions are the worst (such as when accidents occur) should be established in industrial areas of Vietnam.

Increasing Capacity of State Regulatory Agencies: The DOSTEs are the key agencies for implementation of environmental policies in Vietnam. And while DOSTE capacities for monitoring and enforcement have been improved over the last several years, further strengthening is needed. There is also a need for better coordination between DOSTE inspection divisions and environmental management divisions. DOSTEs also need to strengthen their capacities to effectively assign and collect fines and compensations. Mechanisms to increase accountability and reduce the potential for corruption will need to be developed to increase the credibility of the fine system.

Firm Incentives: Firms should be offered training and technical assistance on pollution prevention and cleaner production strategies. Technical assistance programs such as the Vietnam Cleaner Production Center should be expanded. Workshops on waste audit procedures and cost-benefit analysis should be continued. At the same time, charges and fines should be increased and more vigorously applied to motivate pollution prevention.

Community Compensation: Compensation procedures should be strengthened so that community members can demand and receive compensation equivalent to the full damages of pollution impacts. Ultimately this should include the creation of liability laws that establish the right to sue companies and government agencies for environmental malfeasance. This would promote the polluter-pays principle. This might also involve programs for pro bono legal assistance to communities.

Strengthening the Role of NGOs: Formal legal procedures and the rights and roles of domestic and international NGOs need to be clarified and more broadly accepted by the government.

8.7 Advancing Environmental Regulation through Public Information

One clear conclusion from this research is how important information is to pollution debates. Access to information is critical for government officials, factory managers, NGOs, and community members. Good environmental information can support dialogue, transparency, accountability, benchmarking of firms, and public trust in decisions. With access to accurate and understandable environmental information and some capacity building, community members can support a wide range of incentives for pollution reduction. Better educated communities can more effectively evaluate pollution problems and then take constructive steps to pressure firms to reduce pollution. Experiences from other countries, such as the PROPER pollution rating program in Indonesia (Afsah, Laplante, and Wheeler 1995), and the Toxics Release Inventory in the United States (Fung and O'Rourke 2000), show the promise of programs that use environmental information as a tool to achieve pollution reductions.

The example of Indonesia is particularly relevant to Vietnam and other developing countries. Recognizing the many barriers and limitations to enforcing its own environmental laws, the Indonesian Environmental Agency (BAPEDEL) decided in 1995 to create a simple pollution disclosure system called PROPER. The idea of PROPER was to "create incentives for compliance through honor and shame" (Afsah and Ratunanda

1999) by publicly rating the environmental performance of factories. PROPER assigns a color-coded rating—black, red, blue, green, and gold (in order from worst to best)—to factories based on their compliance with environmental standards. Ratings are based on monthly emissions reports filed by firms and corroborated through spot checks by the environmental agency.

This simple, fairly inexpensive program has resulted in significant reductions in pollution since its inception. All of the "black" rated firms improved to red or blue between 1995 and 1997, 46 percent of the red firms improved to blue, and 11 percent of the blue improved to green (Afsah et al. 2000). Factory managers have reported that bad PROPER ratings increased pressure from communities, NGOs and the media, a major motivation for improving environmental performance. Information from PROPER ratings also helped managers make better decisions about pollution reduction as the ratings served as an environmental audit of the factory. The PROPER rating system has also helped to strengthen BAPEDEL, as the agency had to improve the technical capacity and public accountability of its monitoring.

Similar programs supporting public information and new forms of public pressure on firms are being developed in the Philippines, China, Mexico, India, Colombia, Bangladesh, Australia, Mexico, Canada, Denmark, Czech Republic, and Thailand. Public rating programs and "pollutant release and transfer registries" (PRTR) provide information to communities on what they are being exposed to, and create means to use this information to pressure firms and government agencies.

The experience of publishing a "Black Book" of polluting enterprises in Ho Chi Minh City—essentially a list of the worst firms—shows that public disclosure can work in Vietnam as well. This information increased the environmental awareness of the community, created pressures on factories to improve their environmental performance, and provided political support to environmental authorities to force enterprises to comply with environmental standards.

The National Environment Agency in Vietnam is beginning to move towards these kinds of public-information strategies. The NEA (with the support of the World Bank) is currently initiating a pilot project in Hanoi that will rate pollution levels of industrial facilities. This program will develop a simplified process for monitoring, evaluating, classifying, and then rating industrial wastewater from major factories in metropolitan Hanoi. Factories will receive a color-coded rating similar to the Indonesian program that will then be made public. Firms may be given the chance to improve their performance before the rating is made public.

This pollution rating system will likely help promote community-driven processes, and more broadly advance incentives for pollution reduction in Vietnam. If it works, the system will identify the worst polluters in Hanoi, help establish priorities for action, and raise community and firm awareness about pollution problems. Firms will hopefully be motivated to clean up their performance and to do their own monitoring to show they are in compliance. The ratings may also support other social and market pressures on firms. For instance, in the future, consumers in Vietnam may become more environmentally aware, and would be interested to know the ratings of the companies from which they buy products. Also, if firms want to export their products to markets with environmentally aware consumers (such as Europe, Japan, and the United States), the ratings would motivate improved performance. The rating system will also support existing regulatory efforts by generating information for inspectors. The data that goes into determining ratings will support conventional regulatory efforts by identifying the worst polluters in a region, and providing a system for prioritizing regulatory action. The list of "black" firms for instance, could serve as a starting point for inspections.

For these programs to be effective in Vietnam, however, a number of basic policies will need to be developed. The government would need to commit to the public's "right to know" about environmental problems, establish clear procedures for making this information public, and develop procedures for public input into government and firm decision-making.

8.8 Concluding Thoughts

While former state-socialist countries have often been characterized as rife with environmental problems, the challenge for Vietnam is not just to shake off this past, but rather to learn from, and build on, the strengths of its history. If Vietnam can strengthen local autonomy and government responsiveness to environmental and social problems, and at the same time support the existing energies of local community members, the country may yet be able to chart a less destructive path towards industrialization.

The key for Vietnam, as for other countries, is to learn from existing processes which are already working to achieve environmental protections within a challenging context. Understanding what works, when, and how is an important first step to building improved processes. Of course, it is also critical to learn from failures of environmental protection and to understand the barriers and impediments to regulation. Based on rigorous analysis, Vietnamese policy-makers and foreign advisors could begin to extend the successes of Community-Driven Regulation through intentional policy design, planning procedures, technical assistance, and community organizing.

Perhaps above all else, the findings of this research show that community participation can support multiple mechanisms for environmental protection. Community members can play surprisingly intelligent and powerful roles in gathering, analyzing, and deploying information, and then influence decisions which cause environmental harms. Community actions can strengthen existing command-and-control programs, motivate different state roles, promote new forms of pressures through "reputational incentives," support negotiations between firms and communities, and generally make communities more active participants in development decisions and regulations.

But as I have also tried to make clear, because Community-Driven Regulation has a number of weaknesses, baseline standards, monitoring, and reporting are still critical and must be strengthened. In my view, command-and-control strategies are necessary but rarely sufficient in countries like Vietnam. So while the government needs to strengthen its capacity in these areas, it also needs to focus more attention on promoting preventative and participatory strategies in industrial environmental issues. The Vietnamese government can significantly strengthen community participation and transparency in decision making. Development processes that are more open will ultimately also be more credible and publicly acceptable.

The challenges of regulating industry in the global economy can hardly be overstated. But with the energies and commitments of communities, NGOs, the media, and committed government officials, there are effective ways to strengthen the hand of regulation against polluting firms.

Communities can monitor locally and mobilize extralocally, and they can motivate media and NGO scrutiny to magnify public pressures. If coordinated, these actors can help block degrading development while simultaneously helping to chart a course towards more ecologically sustainable development.

There are, of course, many remaining questions and room for further exploration and analysis of these issues. However, existing cases point to important innovations in Vietnam that represent something both new and exciting for future efforts to regulate industrial activities. My hope is that by analyzing when and how these processes of Community-Driven Regulation are effective, it may be possible to advance theories—and ultimately policies—to support more cologically sound and socially just development internationally.

Appendix

Law on Environmental Protection

National Assembly Socialist Republic of Vietnam
Independence-Freedom-Happiness
Hanoi, 27 December 1993

The environment is of special importance to the life of humans and other living creatures as well as to the economic, cultural and social development of the country, the nation, and mankind as a whole.

In order to raise the effectiveness of state management and the responsibilities of the administration at all levels, of state agencies, economic and social organisations, units of the People's Armed Forces and all individuals with respect to environmental protection with a view to protecting the health of the people, ensuring the right of everyone to live in a healthy environment and serving the cause of sustainable development of the country, thus contributing to the protection of regional and global environment;

Pursuant to Article 29 and Article 84 of the 1992 Constitution of the Socialist Republic of Vietnam;

This law provides for the protection of the environment.

Chapter I
General Provisions

Article 1

The environment comprises closely inter-related natural factors and man-made material factors that surround human beings and affect life, production, the existence and development of man and nature.

Environmental protection as stipulated in this law includes activities aimed at preserving a healthy, clean and beautiful environment, improving the environment, ensuring ecological balance, preventing and overcoming adverse impacts of man and nature on the environment, making a rational and economical exploitation and utilisation of natural resources.

Article 2

In this law the below-cited terms shall have the following meanings:

1. *Components of the environment* mean factors that constitute the environment: air, water, soil, sound, light, the earth's interior, mountains, forests, rivers, lakes, sea, living organisms, ecosystems, population areas, production centres, nature reserves, natural landscapes, famed beauty spots, historical vestiges and other physical forms.

2. *Wastes* mean substances discharged from daily life, production processes or other activities. Wastes may take a solid, gaseous, liquid or other forms.

3. *Pollutants* mean factors that render the environment noxious.

4. *Environmental pollution* means alternation in the properties of the environment, violating environmental standards.

5. *Environmental degradation* means qualitative and quantitative alteration in the components of the environment, adversely affecting man's life and nature.

6. *Environmental incidents* mean events or mishaps occurring in the process of human activities, or abnormal changes of nature causing serious environmental degradation. Environmental incidents may be caused by:

a. Storms, floods, droughts, earth cracks, earthquakes, landslides, ground subsidence, volcanic eruptions, acid rain, hails, climatic changes and other natural calamities;

b. Fires, forest fires, technical failures at production or business establishments or in economic, scientific, technical, cultural, social, security or defence facilities, causing damage to the environment;

c. Accidents in the prospection, exploration, exploitation or transportation of minerals or oil and gas, pit collapse, oil spouts and spills, pipeline breaks, shipwrecks, accidents at oil refineries and other industrial establishments;

d. Accidents in nuclear reactors, atomic power plants, nuclear fuel producing or re-processing plants or radioactive material storages.

7. *Environmental standards* mean norms and permissible limits set forth to serve as a basis for the management of the environment.

8. *Clean technology* means a technological process or technical solution either causing no environmental pollution or generating pollutants at the lowest level.

9. *Ecosystem* means a system of groups of living organisms existing and developing together in a given environment, interacting with one another and with that environment.

10. *Biodiversity* means the abundance in gene pools, species and varieties of living organisms and ecosystems in nature.

11. *Environmental impact assessment (E.I.A.)* means the process of analysing, evaluating and forecasting the effects on the environment by socio-economic development projects and plans, by production and business establishments, and economic, scientific, technical, medical, cultural, social, security, defence or other facilities, and proposing appropriate solutions to protect the environment.

Article 3

The State shall exercise unified management of environmental protection throughout the country, draw up plans for environmental protection, build up capabilities for environmental protection activities at the central and local levels.

The State shall adopt investment policies to encourage organisations and individuals at home and abroad to invest under different forms in, and apply scientific and technological advances to, environmental protection, and protect their lawful interests therein.

Article 4

The State shall be responsible for organising the implementation of education, training, scientific and technological research activities and the dissemination of scientific and legal knowledge on environmental protection.

Organizations and individuals shall be liable for participating in the activities mentioned in this Article.

Article 5

The State shall protect national interests with regard to natural resources and the environment.

The State of Vietnam shall broaden cooperative relations with other countries in the world, with foreign organisations and individuals in the field of environmental protection.

Article 6

Environmental protection is the common cause of the entire population.

All organisations and individuals shall have the responsibility to protect the environment, observe the environmental protection legislation, have the right and obligation to detect and denounce any act in breach of the environmental protection legislation.

All foreign organisations and individuals operating on Vietnamese territory shall abide by Vietnam's environmental protection legislation.

Article 7

Organisations and individuals making use of components of the environment for production or business purposes shall, if necessary, contribute financially to environmental protection.

The Government shall regulate the circumstances, levels and modalities for the financial contribution mentioned in this Article.

Any organisation or individual whose activities cause damage to the environment shall make compensation therefore according to regulations by the law.

Article 8

The National Assembly, the People's Councils, the Vietnam Fatherland Front and its member organisations, within the scope of their tasks and powers, shall be responsible for the control and supervision of the implementation of the environmental protection legislation.

The Government and the People's Committees at all levels shall be responsible for organising the implementation of the environmental protection legislation.

Article 9

All acts causing environmental degradation, environmental pollution or environmental incidents, are strictly prohibited.

Chapter II
Prevention and Combat against Environmental Degradation Environmental Pollution and Environmental Incidents

Article 10

The State offices, within the scope of their functions and tasks, shall be responsible for organising the investigation, study and evaluation of the existing conditions of the environment, periodically reporting to the National Assembly on the current status of the environment; for identifying areas of environmental pollution and notifying the public thereof and for drawing up plans for the prevention and combat against environmental degradation, environmental pollution and environmental incidents. Organisations and individuals shall have the responsibility to engage in the prevention and combat against environmental degradation, environmental pollution and environmental incidents.

Article 11

The State encourages, and shall create favourable conditions for all organisations and individuals in the rational use and exploitation of components of the environment, the application of advanced technology and clean technology, the exhaustive use of wastes, the economical use of raw materials and the utilization of renewable energy and biological products in scientific research, production and consumption.

Article 12

Organisations and individuals shall have the responsibility to protect all varieties and species of wild plants and animals, maintain biodiversity and protect forests, seas and all ecosystems.

The exploitation of biological resources must observe their prescribed seasonal characteristics and areas, using proper methods and permitted tools and means in order to ensure their restoration in terms of density, varieties and species, thus preventing ecological imbalance.

The exploitation of forests must comply strictly with plans and specific stipulations of the Law on Forest Protection and Development. The State shall adopt plans to involve organisations and individuals in afforestation and greening of waste lands and denuded hills and mountains to quickly expand the forest cover and protect catchment regions of watercourses.

Article 13

The use and exploitation of nature reserves and natural landscapes must be subject to permission by the sectoral management authority concerned and the State management agency for environmental protection and must be registered with the local People's Committees entrusted with the administrative management of these conservation sites.

Article 14

The exploitation of agricultural land, forest land, and land for aquaculture must comply with land use plans, land improvement plans and ensure ecological balance. The use of chemicals, chemical fertilisers, pesticides and other biological products must comply with stipulations by law.

In carrying out production and business activities or construction works, measures must be taken to restrict, prevent and combat soil erosion, land subsidence, landslide, soil salination or sulphatation, uncontrolled desalination, laterisation and desertification of land, or its transformation into swamps.

Article 15

Organisations and individuals must protect water sources, water supply and drainage systems, vegetation, sanitation facilities, and observe the regulations on public hygiene in cities, urban areas, countryside, population centres, tourism centres and production areas.

Article 16

In carrying out production, business and other activities, all organisations and individuals must implement measures for environmental sanitation and have appropriate waste treatment equipment to ensure compliance with environmental standards and to prevent and combat environmental degradation environmental pollution and environmental incidents.

The Government shall stipulate the nomenclature environmental standards and delegate the authority at different levels for promulgating and supervising the implementation of such standards.

Article 17

Organisations and individuals in charge of the management of economic, scientific, technical, health, cultural, social, security and defence establishments that have begun operation prior to the promulgation of this law must submit an E.I.A. report on their respective establishments for appraisal by the State management agency for environmental protection.

In case of failure to meet environmental standards, the organisations or individuals concerned must take remedial measures within a given period of time as stipulated by the State management agency for environmental protection. Upon expiry of the stipulated time limit, if they still. fail to meet the requirements of the State management agency for environmental protection, the latter shall report to the higher State authority at the next level to consider and decide on the suspension of operation or other penalizing measures.

Article 18

Organisations, individuals when constructing, renovating production areas, population centres or economic, scientific, technical, health, cultural, social, security and defence facilities; owners of foreign investment or joint venture projects, and owners of other socio-economic development projects, must submit E.I.A. reports—to the State management agency for environmental protection for appraisal.

The result of the appraisal of E.I.A. reports shall constitute one of the bases for competent authorities to approve the projects or authorize their implementation. The Government shall stipulate in detail the formats for the preparation and appraisal of E.I.A. reports and shall issue specific regulations with regard to special security and defence establishments mentioned in Article 17 and in this article.

The National Assembly shall consider and make decision on projects with major environmental impacts. A schedule of such types of projects shall be determined by the Standing Committee of the National Assembly.

Article 19

The importation and exportation of technologies, machinery, equipment, biological or chemical products, toxic substances, radioactive materials, various species of animals, plants, gene sources and microorganisms relating to the protection of the environment must be subject to approval by the sectoral management agency concerned and the State management agency for environmental protection.

The Government shall stipulate a schedule for each domain and each category referred to in this Article.

Article 20

While searching, exploring, exploiting, transporting, processing, storing minerals and mineral products, including underground water, organisations and individuals must apply appropriate technology and implement environmental protection measures to ensure that environmental standards are met.

Article 21

While searching exploring, exploiting, transporting, processing, storing oil and-gas, organisations and individuals must apply appropriate technology, implement environmental protection measures, develop preventive plans against oil leakage, oil spills, oil fires and explosions, and disposal facilities necessary to respond rapidly to those incidents.

The use of toxic chemicals in the process of searching, exploration, exploitation, and processing of oil and gas must be guaranteed by technical certificates and be subject to the control and supervision by the State management agency for environmental protection.

Article 22

Organisations, individuals operating means of water, air, road and rail transports must observe environmental standards and be subject to the supervision and periodic inspection for compliance with environmental standards by the relevant sectoral management agency and the State management agency for environmental protection. The operation of transport means failing to meet stipulated environmental standards shall not be permitted.

Article 23

Organisations, individuals producing, transporting, trading, using, storing or disposing of toxic substances, inflammable or explosive substances, must comply with regulations on safety for human and other living beings and must avoid causing environmental degradation, pollution or incidents.

The Government shall stipulate a list of toxic, inflammable or explosive substances mentioned in this Article.

Article 24

The siting, design, construction and operation of plants in the nuclear industry, of nuclear reactors, facilities for nuclear research, for the production, transportation, utilisation and storage of radioactive material, for the disposal of radioactive wastes must comply with legal provisions on nuclear safety and radiation safety and with regulations by the State management agency for environmental protection.

Article 25

Organisations, individuals making use of machinery, equipment, materials with harmful electro-magnetic radiation or ionising radiation must comply with legal provisions on radiation safety and must carry out regular checks and environmental impact assessments of their facilities and report periodically to the State management agency for environmental protection.

Article 26

The choice of sites for collecting, dumping and treating refuse or pollutants and their transportation must comply with regulations by the State management agency for environmental protection and by the local authorities concerned.

Wastewater, refuse containing toxic substances, pathogenetic agents, inflammable or explosive substances, non-degradable wastes, must be properly treated before discharge. The State management agency for environmental protection shall stipulate a schedule of wastewater and refuse mentioned in this Article and supervise their treatment process before discharge.

Article 27

The burial, lying in state, embalmment, interring, cremation and transport of corpses or remains of the dead must utilise progressive methods and means and comply with provisions of the Law on Protection of Public Health to ensure environmental hygiene.

The Administration at all levels must plan for burial, cremation sites and guide people to gradually abandon backward practices.

Cemeteries, crematoria must be located far away from population areas and sources of water.

Article 28

Organisations, individuals in the course of their activities must not cause noises or vibrations that exceed permissible limits, harming the health of surrounding people and adversely affecting their life.

The People's Committees at all levels shall be responsible for the implementation of noise control measures in areas of hospitals, schools, public offices, and residential quarters.

The Government shall promulgate regulations to restrict, and to proceed towards the strict prohibition of the production and firing of firecrackers.

Article 29

The following activities are strictly prohibited:

1. Burning and destruction of forests, uncontrolled exploitation of minerals leading to environmental damage, destroying ecological balance;
2. Discharge of smoke, dust, noxious gas, bad odours causing harm to the atmosphere; emission of radiation, radioactivity exceeding permissible limits into the surrounding environment;
3. Discharge of grease or oil, toxic chemicals, radioactive substances exceeding permissible limits, wastes, dead animals or plants, harmful and infective bacteria and viruses into water sources;
4. Burial, discharge of toxic substances exceeding permissible limits into the soil;

5. Exploitation, trading in precious or rare species of plants and animals identified in the schedule stipulated by the Government;

6. Importation of technology and equipment not meeting environmental standards; importation, exportation of wastes;

7. Use of methods, means, instruments causing massive destruction in exploiting or harvesting animal and plant resources.

Chapter III
Remedy of Environmental Degradation, Environmental Pollution, Environmental Incidents

Article 30

Organisations, individuals engaged in production, business and other activities that cause environmental degradation, environmental pollution, or environmental incidents must implement remedial measures as specified by the local People's Committees and by the State management agency for environmental protection, and shall be liable for damages according to regulations by the law.

Article 31

Organisations, individuals allowing radioactivity, electro-magnetic radiation, ionising radiation to exceed permissible limits must take immediate measures to control and remedy the consequences, timely report to the relevant sectoral management agency and to the State management agency for environmental protection, as well as to the local People's Committee to resolve the problem.

Article 32

The remedy of an environmental incident includes: eliminating the cause of the incident; rescuing people and property; assisting, stabilising the life of the people; repairing damaged facilities; restoring production; sanitising the environment, preventing and combating epidemics; investigating, collecting statistics on damages, monitoring changes to the environment; and rehabilitating the environment of the affected area.

Article 33

Persons who detect signs of an environmental incident must immediately notify the local People's Committee, the nearest agency or organisation for timely action:

Organisations, individuals at the site of the environmental incident must take appropriate measures to timely remedy it and immediately report to the superior administrative authority, the nearest People's Committee and the State management agency for environmental protection.

Article 34

The chairman of the People's Committee of the locality where the environmental incident occurs is empowered to order an emergency mobilization of manpower, materials and other means for remedial actions.

If the environmental incident occurs in an area covering several localities, the Chairmen of the respective local People's Committees shall cooperate to take remedial actions.

In case the incident is beyond local remedy capability, the Minister of Science, Technology and environment in conjunction with the heads of the agencies concerned shall determine the application of remedial measures and report to the Prime Minister.

Article 35

In case the environmental incident is of special severity, the Prime Minister shall determine the application of urgent remedial measures.

When such incident has been brought under control the Prime Minister shall determine the revocation of the application of the urgent remedial measures.

Article 36

The agencies that are empowered to mobilise manpower, materials, and other means to remedy environmental incidents must reimburse the mobilised organisations, individuals for their expenses according to regulations by the law.

Chapter IV
State Management of Environmental Protection

Article 37

The scope of State management of environmental protection includes:

1. Promulgating, and organising the implementation of, statutory instruments on environmental protection; promulgating systems of environmental standards;
2. Developing, and guiding the implementation of, strategies and policies of environmental protection, plans to prevent, control and remedy environmental degradation, environmental pollution, environmental incidents;
3. Establishing and managing environmental protection facilities, and facilities relating to environmental protection;
4. Organising, establishing and managing monitoring systems, periodically assessing the current state of the environment, forecasting environmental changes;
5. Appraising E.I.A. reports on projects and on production or business establishments;
6. Issuing, revoking certificates of compliance with environmental standards;

7. Supervising, inspecting, checking the observance of environmental protection legislation; settling disputes, appeals or complaints concerning environmental protection; dealing with breaches of environmental protection legislation;
8. Training personnel in environmental science and management; educating, propagandising, disseminating knowledge and legislation in environmental protection;
9. Organising research and development activities and application of scientific and technological advances in the field of environmental protection;
10. Developing international relations in the field of environmental protection.

Article 38

The Government shall, pursuant to its power and responsibility, exercise unified State management of environmental protection throughout the country.

The Ministry of Science, Technology and Environment shall be responsible to the Government for exercising the function of State management of environmental protection.

All ministries, ministry-level agencies and other Government bodies shall, within the scope of their respective functions, powers and responsibilities, cooperate with the Ministry of Science, Technology and Environment in carrying out environmental protection within their sectors and in establishments under their direct supervision.

The People's Committees of provinces and cities directly under the Central Government shall exercise their State management function for environmental protection at the local level.

The Services of Science, Technology and Environment shall be responsible to the People's Committees of provinces and cities directly under the Central Government, for environmental protection in their localities.

Article 39

The system of organisation, functions, responsibilities and powers of the State management agency for environmental protection shall be determined by the Government.

Article 40

The State management agency for environmental protection shall carry out its function of specialized inspection on environmental protection and be responsible to coordinate with specialized inspectors of the ministries and sectors concerned in the protection of the environment.

The organisation, obligations, powers, activities and coordination of specialized inspectors in the protection of the environment shall be determined by the Government.

Article 41

During the inspection process, the Inspection Team or Inspector is empowered to:

1. Require the organisations, individuals concerned to provide documents and reply to questions on matters necessary for inspection;
2. Conduct technical control measures on-site;
3. Decide to temporarily suspend, in case of emergency, activities which threaten to cause serious environmental incidents and be responsible for such decision before the law, and at the same time, immediately report the case to the competent State agency for decision or recommend the latter to suspend activities likely to cause environmental incidents;
4. Deal within their competence or recommend the competent State agency to deal with breaches of the law.

Article 42

Organisations, individuals must create favourable conditions for the Inspection Team or the Inspector to carry out their functions and must observe the decisions of the Inspection Team or the Inspector.

Article 43

Organisations, individuals are entitled to appeal to the Head of the agency which decides the inspection against the conclusions and decisions adopted by the Inspection Team or the Inspector with regard to their establishment.

Organisations, individuals have the right to complain, denounce to the State management agency for environmental protection or other competent State agencies about activities in breach of environmental protection legislation.

Agencies receiving complaints, denunciations shall be responsible for their examination and resolution in accordance with regulations by the law.

Article 44

In case there are several organisations, individuals operating within an area where environmental incidents, environmental pollution or environmental degradation occur, the power to determine the responsibility assigned to those organisations, individuals for remedial measures is defined as follows.

1. For environmental incidents, environmental pollution or environmental degradation occurring within a province or a city directly under the Central Government, the responsible parties shall be determined by the specialized environmental protection inspector of that province, city, or proposed and reported by the latter to the Chairman of the People's Committee of that province or city for consideration and decision. If one or more parties disagree with that decision, they shall be entitled to appeal to the Minister of Science, Technology and Environment. The decision of the Minister of Science, Technology and Environment shall prevail.
2. For environmental incidents, environmental pollution or environmental degradation occurring in two or more provinces, or cities directly under the Central

Government, the responsible parties shall be determined by the specialized environmental protection inspector of the Ministry of Science, Technology and Environment or proposed and reported by the latter to the Minister of Science, Technology and Environment for consideration and decision. If one or more parties disagree with the decision of the Minister of Science, Technology and Environment, they shall be entitled to appeal to the Prime Minister for decision.

Chapter V
International Relations with Respect to Environmental Protection

Article 45

The Government of Vietnam shall implement all international treaties and conventions relating to the environment which it has signed or participated in, honour all international treaties and conventions on environmental protection on the basis of mutual respect for each other's independence, sovereignty, territorial integrity and interests.

Article 46

The Government of Vietnam adopts priority policies towards countries, international organisations, foreign organisations and individuals with respect to environmental manpower training, environmental scientific research, clean technology application, development and implementation of projects for environmental improvement, control of environmental incidents, environmental pollution, environmental degradation, and projects for wastes treatment in Vietnam.

Article 47

Organizations, individuals and owners of transportation means which, in transit through the Vietnamese territory, carry potential sources of environmental incidents or environmental pollution must apply for permission, declare and submit to the control and supervision by the State management agency for environmental protection of Vietnam. Any breach of Vietnamese environmental protection legislation shall, depending on the extent of the infringement, be dealt with according to Vietnamese law

Article 48

Any dispute concerning environmental protection on the Vietnamese territory in which one or all parties are foreigners shall be settled according to Vietnamese law, taking into account international laws and practices.

Any dispute between Vietnam and other countries in the field of environmental protection shall be settled on the basis of negotiation, taking into account international laws and practices.

Chapter VI
Rewards and Dealing with Breaches

Article 49

Organisations, individuals having good records in environmental protection activities, in the early detection and timely report of signs of environmental incidents, in the remedy of environmental incidents, environmental pollution, environmental degradation, in the prevention of acts which damage the environment shall be rewarded. Those who suffer damage to their property, health or life, while participating in the protection of the environment, in the remedy of environmental incidents, environmental pollution, environmental degradation and in the combat against activities violating environmental protection legislation, shall be compensated according to regulations by the law.

Article 50

Those who commit acts of destruction or cause damage to the environment, who disregard the order of mobilisation by the competent State agency upon the occurrence of environmental incidents, who fail to implement regulations on environmental impact assessment, or infringe other legal provisions for environmental protection shall be dealt with administratively or be criminally prosecuted, depending on the nature and extent of the infringement and the consequences.

Article 51

Those who take advantage of their positions and powers to infringe environmental protection legislation, to protect persons infringing the environmental protection legislation, whose lack of responsibility allows environmental incidents or environmental pollution to occur, shall be disciplined or be criminally prosecuted, depending on the nature and extent of the infringement and the consequences.

Article 52

Organisations, individuals that commit acts of violation against the environmental protection legislation, causing damage to the State, to other organisations or individuals, shall, in addition to the penalties specified in Article 50 and 51 of this Law, compensate for the damages and costs of remedying the consequences, according to regulations by the law.

Chapter VII
Implementation Provisions

Article 53

Domestic or foreign organisations, individuals that have caused serious damage to the environment prior to the promulgation of this Law, with long-term adverse impacts on the environment and the health of the people shall, depending on the extent of the consequences, be liable for the damages and the rehabilitation of the environment, according to regulations by the Government.

Article 54

This Law shall take effect from the date of its promulgation.
   All previous stipulations which contradict this Law are revoked.

Article 55

The Government shall regulate in detail the implementation of this Law.
   This Law was passed on 27 December 1993 by the National Assembly of the Socialist Republic of Vietnam, 9th Legislature, at its 4th Session.

Chairman of the National Assembly

Signed: Nong Duc Manh
(Published by the National Political Publishing House, July 1994)

Notes

Chapter 1

1. As Giang (2001) notes, Vietnam "has just four state employees in the field for every million people compared with Malaysia's 100, Cambodia's 55, Thailand's 30 and China's 20."

2. See for instance, UNDP (1999) "A study on aid to the environment sector in Vietnam."

3. These included a wide variety of firms and communities ranging from a Coca-Cola plant in a rural area, to a rubber factory in an urban area, from large multinationals, to medium-sized Vietnamese state-owned enterprises, to small co-operative enterprises. The pool of 120 cases was narrowed to thirty cases, all of which I visited and analyzed in some detail. For these thirty cases, I conducted factory tours, reviewed EIAs and waste audits, and completed short interviews with a sample of factory managers and community members. Analysis of these 120 cases ultimately also helped to contextualize the six detailed cases.

Chapter 2

1. *Doi Moi* literally means "renovation" in Vietnamese. However, the phrase has come to represent a broad set of economic and political reforms begun in 1986.

2. As this book went to press a new ministry was created to handle environmental issues—the Ministry of Natural Resources and Environment (MONRE). MONRE has assumed many of the roles and responsibilites of MOSTE. Throughout this book we refer to MOSTE, as it was the agency responsible for environmental issues at the time.

Chapter 3

1. Approximately 78 percent of Vietnamese report they belong to no religion, while around 7 percent are Catholic.

2. While Vietnam remains a one-party Communist state, the government has initiated democratic reforms including local-level elections of People's Council representatives and National Assembly members. To date, little competition exists in these elections; however, the scope and power of elected officials (particularly National Assembly members) appears to be increasing.

3. McAdam et al. (1996) define mobilizing structures as "those collective vehicles, informal as well as formal, through which people mobilize and engage in collective action," and in particular "organizations and informal networks that comprise the collective building blocks of social movements."

4. Televised debates of the National Assembly have recently shown surprising levels of candor and criticism.

5. I reviewed letters submitted to the Dong Nai and Phu Tho DOSTEs.

Chapter 5

1. It should be noted that these institutional arrangements continue to evolve in Vietnam.

2. While there is no mass environmental "movement" in Vietnam, Vietnamese press reports and my own interviews indicate that complaints and protests regarding environmental issues have become more widespread over the last five years.

Chapter 6

1. Personal communication with Thuyen Nguyen, director of Vietnam Labor Watch, June 30, 1998.

2. Nike reports that Tae Kwang Vina reduced work hours in this factory after this research was completed.

3. Leading environmental NGOs include the Center for Natural Resources and Environmental Studies (CRES), the Institute for Environmental Science and Technology (INEST), the Center for Training and Research on Water Supply and Environment (CEFINEA), the Center for Environmental Engineering of Towns and Industrial Areas (CEETIA), the Center for Environmental Technology and Management (CENTEMA), and the Environmental Protection Center (EPC).

4. The Ernst & Young report and my analysis of it are available on the internet at: <http://www.daraorourke.net>.

5. In a scene from *The Big One*, 1997, Michael Moore discusses child labor with Phil Knight the CEO of Nike. Moore asks: "Twelve-year-olds working in [Indonesian] factories? That's okay with you?" Knight: "They're not twelve-year-olds working in factories . . . the minimum age is fourteen." Moore: "How about fourteen then? Does that bother you?" Knight: "No."

6. These headlines came from both Vietnamese and English language papers during 1997 and 1998.

7. As mentioned in chapter 4, the editor of a business paper was jailed in 1999 for reporting on a fairly clear case of corruption in a customs office. He was ultimately exonerated, but not before spending significant time in jail, which served as a very strong message about the potential costs of reporting on these issues.

Chapter 7

1. There is one recent exception to this pattern. The National Environment Agency instigated a nationwide program in 1997 to inspect firms for compliance with the EIA law. This program was simply an attempt to assess how many firms had conducted EIAs, and whether firms were complying with their EIAs.

2. McCubbins and Schwartz (1984) discuss models of congressional oversight, which they call "patrol" and "alarm" models. In the patrol model, Congress establishes systems to regularly evaluate implementation of legislation. Under the alarm model, Congress creates systems that trigger action such as monitoring by Congress.

3. My use of the term "fire alarm" should be distinguished from McCubbins and Schwartz's (1984) use. In my cases, the alarms are pulled by community members, not government personnel. The alarm metaphor is meant to describe a relationship between the community and the state that moves beyond traditional command-and-control regulation, and in which the community has a role in driving the regulatory process.

4. See Rueschemeyer and Evans (1985: 48) for a discussion of the tendency of states to be simultaneously corporate actors and arenas in which conflicts are played out.

5. The Vietnamese government (with the assistance of Sweden, Canada, and other external supporters) is working to build these environmental capacities. Staff are being trained to perform ambient and point-source sampling, review EIAs, issue permits, develop compliance schedules, set and collect fines, and write reports. The professionalization of environmental duties will likely strengthen the actions of state agencies as staff begin to believe more in their responsibilities, or at least work to avoid being labeled as incompetent or corrupt. The fruits of these efforts, however, had not been realized at the time of my research.

6. As Ferguson (1998) points out, states should be viewed "not in opposition to something called society, but as themselves composed of bundles of social practices, every bit as local in their social situatedness and materiality."

References

Aden, Jean, Ahn Kyu-Hong, and Mike Rock. 1999. What is Driving the Pollution Abatement Expenditure Behavior of Manufacturing Plants in Korea? *World Development* 27 (7): 1203–1214.

Afsah, Shakeb, Allen Blackman, and Damayanti Ratunanda. 2000. How Do Public Disclosure Pollution Control Programs Work? Evidence from Indonesia. Resources for the Future, Discussion Paper No. 00-44.

Afsah, Shakeb, Benoit Laplante, and David Wheeler. 1995. What is PROPER? Reputational Incentives for Pollution Control in Indonesia. Policy Research Department, Washington, D.C.: World Bank.

Afsah, Shakeb, Benoit Laplante, and David Wheeler. 1996. *Controlling Industrial Pollution: A New Paradigm*. Washington, D.C.: World Bank.

Afsah, Shakeb, and Damayanti Ratunanda. 1999. Environmental Performance Measurement and Reporting in Developing Countries: The Case of Indonesia's Program for Pollution Control Evaluation and Rating (PROPER). In *Sustainable Measures: Evaluation and Reporting of Environmental and Social Performance*, eds. M. Bennett and P. J. Sheffield, 185–201. London: Greenleaf Publishing.

Agarwal, Bina. 2000. Conceptualizing Environmental Collective Action: Why Gender Matters. *Cambridge Journal of Economics* 24: 283–310.

Agthe, Donald E., R. Bruce Billings, and James Marchand. 1996. Socioeconomic and Political Determinants of State Spending on Environmental Programs. *American Economist* 40 (1): 24.

Amsden, Alice. 1989. *Asia's Next Giant—South Korea and Late Industrialization*. New York: Oxford University Press.

Amsden, Alice. 1991. Third World Industrialization: Global Fordism or a New Model? *New Left Review* (182): 5–31.

Andreff, Wladimir. 1993. The Double Transition from Underdevelopment and from Socialism in Vietnam. *Journal of Contemporary Asia* 23 (4): 515–531.

Andrews, Richard. 1992. Environmental Policy-Making in the United States. Paper presented at the conference *Towards a Transatlantic Environmental Policy*. Washington, D.C., Jan. 9–10.

Andrews, Richard. 1999. *Managing the Environment, Managing Ourselves—A History of American Environmental Policy.* New Haven: Yale University Press.

Ashford, Nicholas A. 2000. An Innovation-Based Strategy for a Sustainable Environment. In *Innovation-Oriented Environmental Regulation,* ed. J. Hemmelskamp, K. Rennings, and F. Leone, 67–107. New York: Springer-Verlag.

Ashford, Nicholas A., and George Heaton. 1983. Regulation and Technological Innovation in the Chemicals Industry. *Law and Contemporary Problems* 46 (3): 109–157.

Asia Pulse. 2001. *Vietnam Re-evaluates Last Year's Major Socioeconomic Targets.* Hanoi: Asia Pulse.

Asia Times. 2000. Environmental Crisis Takes Over Vietnam's Economic Zones. February 24, 2000. Accessed on the Web at <http://atimes.com/se-asia/BB24Ae01.html>.

Associated Press. 2002. Angered by Pollution, Villagers Damage a Factory in Vietnam. Wednesday, October 23, 2002. Accessed on the Web at <www.enn.com/news/wire-stories/2002/10/10232002/ap_48782.asp>.

Avery, Christopher. 1999. Business and Human Rights in a Time of Change. In *Democracy, Human Rights and the Rule of Law,* ed. V. Iyer. New Delhi: Butterworth.

Barber, Charles Victor. 1997. The Case Study of Indonesia. In *The AAAS Project on Environmental Scarcities, State Capacity, and Civil Violence.* Washington, D.C.: Committee on International Security Studies.

Barnet, Richard, and John Cavanagh. 1994. *Global Dreams: Imperial Corporations and the New World Order.* New York: Simon & Schuster.

Bartlett, Robert. 1994. Evaluating Environmental Policy Success and Failure. In *Environmental Policy in the 1990s,* ed. N. Vig and M. Kraft, 167–187. Washington, D.C.: CQ Press.

Beck, Ulrich. 1992. *Risk Society.* Beverly Hills, Calif.: Sage.

Bell, Colin, and Howard Newby. 1972. *Community Studies: An Introduction to the Sociology of the Local Community.* New York: Praeger.

Bello, Walden, and Stephanie Rosenfeld. 1990. *Dragons in Distress: Asia's Miracle Economies in Crisis.* San Francisco: Institute for Food and Politics.

Benjamin, Medea. 1999. Nike: What's It All About? Position paper available on the Internet at: <www.globalexchange.com>.

Beresford, Melanie. 1988b. *Vietnam Politics, Economics and Society.* London: Pinter Publishers.

Beresford, Melanie, and Lyn Fraser. 1992. Political Economy of the Environment in Vietnam. *Journal of Contemporary Asia* 22 (1): 3–19.

Bissell, Trim. 1999. Nike, Reebok Compete to Set Labor Rights Pace. Position paper available on the Internet at: <www.compugraph.com/clr>.

Brown, Phil. 1992. Popular Epidemiology and Toxic Waste Contamination: Lay and Professional Ways of Knowing. *Journal of Health and Social Behavior* 33: 267–281.

Buffett, Sandy. 1996. Reconciling Disparate Paradigms: Environmental Sustainability and Economic Growth in Viet Nam. Masters Thesis, American University International Development Program, Washington, D.C. Available on the web at: <http://gurukul.ucc.american.edu/ted/papers/vietnam.htm>.

Bullard, Robert. 1994. *Unequal Protection: Environmental Justice and Communities of Color.* San Francisco: Sierra Club Books.

Burawoy, Michael. 1995. The Extended Case Method: Steering a Course Between Positivism and Postmodernism. Working paper.

Burrows, Paul. 1980. *The Economic Theory of Pollution Control.* Cambridge, Mass.: MIT Press.

Buttel, Frederick H. 1992. Environmentalization: Origins, Processes, and Implications for Rural Social Change. *Rural Sociology* 57 (1): 1–27.

Cairncross, Frances. 1992. *Costing the Earth.* Boston: Harvard Business School Press.

Canter, Larry W. 1996. *Environmental Impact Assessment.* Second Edition. New York: McGraw-Hill.

Carew-Reid, Jeremy. 1998. Assessing Progress in Sustainable Development in Vietnam—A Discussion Paper. Vietnam Capacity 21 Project, Ministry of Planning and Investment, Hanoi.

Chess, Caron. 2000. Improving Public Participation in Solving Environmental Health Problems. *Journal of Environmental Health,* 63 (1): 24–27.

Coase, Ronald. 1960. The Problem of Social Cost. *Journal of Law and Economics* 3 (October): 1–44.

Cohen, Mark. 1998. Monitoring and Enforcement of Environmental Policy. Unpublished manuscript available on the Internet at <www.worldbank.org/nipr/>.

Commoner, Barry. 1990. *Making Peace with the Planet.* New York: Random House.

Communist Party (CP). 1998. On Strengthening Environmental Protection during the Period of National Industrialization and Modernization. Directive No. 36 CT/TW. Hanoi.

Daly, Herman. 1991. *Steady-State Economics.* Washington, D.C.: Island Press.

Daly, Herman, and John Cobb. 1989. *For the Common Good—Redirecting the Economy Toward Community, the Environment, and a Sustainable Future.* Boston: Beacon Press.

Dasgupta, Susmita, Benoit Laplante, N. Maminga, and Hua Wang. 2000. Local Enforcement of Environmental Regulation in China: The Role of Inspection and Complaint. Washington, D.C.: World Bank Policy Research Working Group.

Dasgupta, Susmita, and David Wheeler. 1996. Citizen Complaints as Environmental Indicators: Evidence from China. Washington, D.C.: World Bank.

Department of Planning and Investment (DPI). 2002. Ho Chi Minh City Macroeconomics. Available on the Web at <http://www.hcminvest.gov.vn/html/eco1.html>.

Desai, Uday (ed.). 1998. *Ecological Policy and Politics in Developing Countries.* Albany: State University of New York Press.

Deutsche Presse-Agentur. 2001. Four Sugar Workers Crushed to Death in Vietnam. May 7.

Deyo, Frederic C. (ed.). 1987. *The Political Economy of the New Asian Industrialism.* Ithaca: Cornell University Press.

Di Chiro, Giovanna. 1998. Environmental Justice from the Grassroots. In *The Struggle for Ecological Democracy—Environmental Justice Movements in the U.S.,* ed. D. Faber. New York: Guilford Press.

Dillon, P. S., and K. Fischer. 1992. *Environmental Management in Corporations: Methods and Motivations.* Medford, Mass.: Tufts University Center for Environmental Management.

Dorf, Michael, and Charles Sabel. 1998. A Constitution of Democratic Experimentalism. *Columbia Law Review* 98 (2): 267–473.

Douglass, Mike. 2001. Social Capital and Livable Communities: Urban Poverty and the Environment in Seoul and Bangkok. In *Livable Cities: The Politics of Urban Livelihood and Sustainability,* ed. P. Evans. Berkeley: University of California Press.

Dryzek, John. 1994. *Discursive Democracy: Politics, Policy, and Political Science.* New York: Cambridge University Press.

Duane, Timothy. 1992. Environmental Planning and Policy in a Post-Rio World. *Berkeley Planning Journal* 7: 27–47.

Dwivedi, O. P., and Dhirendra K. Vajpeyi. 1995. *Environmental Policies in the Third World: A Comparative Perspective.* Westport, Conn.: Greenwood.

Economist Intelligence Unit (EIU). 1997. *Vietnam Country Profile.* London: EIU.

Economist Intelligence Unit (EIU). 1998. *Vietnam Country Profile.* London: EIU.

Economist Intelligence Unit (EIU). 2000a. *Vietnam Country Profile 2000,* London.

Economist Intelligence Unit (EIU). 2000b. *Vietnam Country Report,* October 2000, London.

Economist Intelligence Unit (EIU). 2001. *Vietnam Country Profile.* London: EIU.

Economy, Elizabeth. 1997. The Case Study of China: Reforms and Resources. Published as a paper in *AAAS Project on Environmental Scarcities, State Capacity, and Civil Violence,* Washington, D.C.: Committee on International Security Studies.

Eder, Norman. 1996. *Poisoned Prosperity—Development, Modernization, and the Environment in South Korea.* Armonk, N.Y.: M. E. Sharpe.

England, Sara B., and Daniel M. Kammen. 1993. Energy Resources and Development in Vietnam. *Annual Review of Energy and Environment* 18: 137–167.

Evans, Peter. 1995. *Embedded Autonomy—State and Industrial Transformation.* Princeton: Princeton University Press.

Evans, Peter. 1996. Government Action, Social Capital and Development: Creating Synergy Across the Public-Private Divide. *World Development* 24 (6): 1033–1132.

Evans, Peter. 1997. The Eclipse of the State? Reflections on Stateness in an Era of Globalization. *World Politics* 50 (1): 62–87.

Evans, Peter. 2000. Counter-Hegemonic Globalization: Transnational Networks as Political Tools for Fighting Marginalization. *Contemporary Sociology,* 29 (1): 230–241.

Evans, Peter. 2001. *Livable Cities: The Politics of Urban Livelihood and Sustainability.* Berkeley: University of California Press.

Faber, Daniel. (ed.) 1998. *The Struggle for Ecological Democracy: Environmental Justice Movements in the United States.* New York: Guilford Press.

Ferguson, James. 1998. Transnational Topographies of Power: Beyond the State and Civil Society. In the Study of African Politics, unpublished manuscript presented at the Institute for International Studies, UC Berkeley, April 24, 1998.

Fforde, Adam. 1993. The Political Economy of 'Reform' in Vietnam. In *The Challenge of Reform in Indochina,* ed. B. Ljunggren, 293–326. Harvard Institute for International Development.

Fforde, Adam, and Stefan de Vylder. 1988. *Vietnam—An Economy in Transition.* Stockholm: Swedish International Development Authority.

Fforde, Adam, and Stefan de Vylder. 1996. *From Plan to Market—The Economic Transition in Vietnam.* Boulder: Westview Press.

Fiorino, Daniel J. 1995. *Making Environmental Policy.* Berkeley: University of California Press.

Fischer, Kurt, and Schot, Johan. 1993. *Environmental Strategies for Industry: International Perspectives on Research Needs and Policy Implications.* Washington: Island Press.

Flaherty, M., and A. Rappaport. 1991. *Multinational Corporations and the Environment: A Survey of Global Practices.* Medford, Mass.: Tufts University Center for Environmental Management.

Foley, Carol. 1995. Finding the System Boundaries: Changing Goals and Methods for Promoting Pollution Prevention. Paper presented at the Conference Lessons Learned: Using Technical Assistance to Promote Pollution Prevention, June 27–28, Washington, D.C.

Fortmann, Louise, and Emory Roe. 1993. On Really Existing Communities—Organic or Otherwise. *Telos* 95: 139–146.

Fox, Jonathan, and David Brown. (eds.). 1998. *The Struggle for Accountability: the World Bank, NGO's and Grassroots Movements.* Cambridge, Mass.: MIT Press.

Fung, Archon, and Dara O'Rourke. 2000. Reinventing Environmental Regulation from the Grassroots Up: Explaining and Expanding the Success of the Toxics Release Inventory. *Environmental Management* 25 (2): 115–127.

Gamson, William. 1990. *The Strategy of Social Protest.* Second Edition. Belmont, Calif.: Wadsworth Inc.

Garkovich, Lorraine. 1989. *Population and Community in Rural America.* New York: Greenwood.

Gaventa, John. 1980. *Power and Powerlessness—Quiescence and Rebellion in an Appalachian Valley.* Chicago: University of Illinois Press.

General Statistical Office. 1996. *Summary of Rural Census, 1994.* Hanoi: Statistical Publishing House.

General Statistical Office. 1998. *Statistical Yearbook 1997.* Hanoi: Statistical Publishing House.

General Statistical Office. 1999. Available at <http://www.worldbank.org.vn>.

General Statistical Office. 2000. *Figures on Social Development in the 1990s in Vietnam.* Hanoi: Statistical Publishing House.

General Statistical Office. 2001. *Statistical Yearbook 2000.* Hanoi: Statistical Publishing House.

Gerstenzang, James. 2001. Bush Defends His Stance on Environment. *Los Angeles Times,* March 30, 2001, page 1.

Giang, Tu. 2001. Nation Scores Low in "Green Guardian" Stakes. *The Vietnam Investment Review,* November 5, 2001.

Gibson, J., and F. Halter. 1994. Strengthening Environmental Law In Developing Countries. *Environment* (Jan–Feb) 36 (1): 40–43.

Giddens, Anthony. 1990. *The Consequences of Modernity.* Cambridge: Polity Press.

Goldman, Michael. 2000. Constructing an Environmental State: Eco-governmentality and other Transnational Practices of a Green World Bank. Draft manuscript.

Gottlieb, Robert. 1993. *Forcing the Spring: The Transformation of the American Environmental Movement.* Washington, D.C.: Island Press.

Gottlieb, Robert. (ed.). 1995. *Reducing Toxics: A New Approach to Policy and Industrial Decision-making.* Washington, D.C.: Island Press.

Gottlieb, Robert, Maureen Smith, Julie Roque, and Pamela Yates. 1995. New Approaches to Toxics: Production Design, Right-to-Know, and Definition Debates. In *Reducing Toxics: A New Approach to Policy and Industrial Decision Making,* ed. Robert Gottlieb, 124–165. Washington, D.C.: Island Press.

Gould, Kenneth, Allan Schnaiberg, and Adam Weinberg. 1996. *Local Environmental Struggles: Citizen Activism in the Treadmill of Production.* New York: Cambridge University Press.

Greenfield, Gerard. 1994. The Development of Capitalism in Vietnam. *The Socialist Registrar 1994,* 30: 202–234.

Greenhouse, Steven. 1997. Nike Shoe Plant in Vietnam is Called Unsafe for Workers. *New York Times,* Saturday, November 8, p. A1.

Haggard, Stephan. 1990. *Pathways from the Periphery: The Politics of Growth in the Newly Industrializing Countries.* Ithaca, N.Y.: Cornell University Press.

Harvie, Charles, and Tran Van Hoa. 1997. *Vietnam's Reforms and Economic Growth.* London: Macmillan.

Hatfield Consultants. 1998. Preliminary Assessment of Environmental Impacts Related to Spraying of Agent Orange Herbicide During the Viet Nam War. Consultant report, vol. 1 and 2, Vancouver, Canada.

Hayes, Denis. 1991. Harnessing Market Forces to Protect the Earth. *Issues in Science and Technology*, 90–91: 46–51.

Hearne, Shelley. 1996. Tracking Toxics: Chemical Use and the Public's Right-to-Know. *Environment*, 38 (6): 5–33.

Henriques, Irene, and Perry Sadorsky. 1996. The Determinants of an Environmentally Responsive Firm: An Empirical Approach. *Journal of Environmental Economics and Management*, 30: 381–395.

Hettige, Hemamala, Mainul Huq, Sheoli Pargal, and David Wheeler. 1995. Determinants of Pollution Abatement in Developing Countries: Evidence from South and South-East Asia. Policy Research Department, Washington, D.C.: World Bank.

Hillery, G. 1955. Definitions of Community: Areas of Agreement. *Rural Sociology* 20: 111–123.

Hirschorn, Joel, and Kirsten Oldenburg. 1991. Prosperity without Pollution: The Prevention Strategy for Industry and Consumers. New York: Van Nostrand Rheinhold.

Huq, Mainul, and David Wheeler. 1992. Pollution Reduction Without Formal Regulation: Evidence from Bangladesh. Policy Research Department, Washington, D.C.: World Bank.

Interpress Service. 2001. Vietnam: Amid Industrialization, Country Choking on Development. November 22, 2001.

Jamieson, Neil. 1991. Culture and Development in Vietnam. East–West Center Working Paper No. 1, Honolulu.

Jamieson, Neil. 1993. *Understanding Vietnam*. Berkeley: University of California Press.

Jancar-Webster, Barbara. 1993. *Environmental Action in Eastern Europe: Responses to Crises*. New York: M. E. Sharpe.

Japan International Cooperation Agency (JICA). 1998. *Household Survey on Environmental Awareness of Hanoi Citizens*. Prepared for the Study for Environmental Improvement for Hanoi City, Hanoi, November.

Johnston, R. J., D. Gregory, and D. M. Smith. 1994. *The Dictionary of Human Geography*. 3rd edition. Oxford: Blackwell.

Karliner, Joshua. 1997. *The Corporate Planet: Ecology and Politics in the Age of Globalization*. San Francisco: Sierra Club Books.

Keck, Margaret, and Kathryn Sikkink. 1998. *Activists Beyond Borders—Advocacy Networks in International Politics*. Ithaca: Cornell University Press.

Kerkvliet, B. J. T. 1995. Village-State Relations in Vietnam: The Effect of Everyday Politics on Decollectivization. *The Journal of Asian Studies*, 54 (2): 396–418.

Khien, Nguyen Duc. 1996. Report on Environmental Management Practices of Hanoi's DOSTE. Paper presented at the VCEP Seminar on Environmental Management, Hanoi, August 2–23.

King, Gary, Robert Keohane, and Sidney Verba. 1994. *Designing Social Inquiry*. Princeton: Princeton University Press.

Kolko, Gabriel. 1995. Vietnam Since 1975: Winning a War and Losing the Peace. *Journal of Contemporary Asia* 25 (1): 3–49.

Korten, David C. 1995. *When Corporations Rule the World.* Bloomfield, Conn.: Kumarian Press.

Kraft, M. E., and N. J. Vig (eds.). 1990. *Environmental Policy in the 1990s.* Washington, D.C.: Congressional Quarterly.

Laplante, Benoit. 1995. Industrial Pollution Control: Role of Public Information. Presentation to the Economic Development Institute Workshop on Industrial Pollution Prevention in Vietnam, Washington, D.C., Sept. 13th.

Lawrence, Anne T., and David Morell. 1995. Leading-Edge Environmental Management: Motivation, Opportunity, Resources, and Processes. *Research in Corporate Social Performance and Policy,* Suppl. 1: 99–126.

Lester, James. 1995. *Environmental Politics and Policy—Theories and Evidence.* Durham: Duke University Press.

Lipshutz, Ronnie. 1996. *Global Civil Society and Global Environmental Governance.* Albany: State University of New York Press.

Ljunggren, Börje. (ed.). 1993. *The Challenge of Reform in Indochina.* Cambridge, Mass.: Harvard Institute for International Development.

Luong, Hy Van. 1992. *Revolution in the Village—Tradition and Transformation in North Vietnam, 1925–1988.* Honolulu: University of Hawaii Press.

Mander, Jerry and Edward Goldsmith (eds.). 1996. *The Case against the Global Economy.* San Francisco: Sierra Club Books.

Marr, David, and Christine White. (eds.). 1988. *Postwar Vietnam: Dilemmas in Socialist Development.* Cornell Southeast Asia Program.

McAdam, Doug, McCarthy, John, and Mayer Zald. 1996. *Comparative Perspectives on Social Movements.* New York: Cambridge University Press.

McCubbins, Mathew D., and Thomas Schwartz. 1984. Congressional Oversight Overlooked: Police Patrols versus Fire Alarms. *American Journal of Political Science* 28: 165–179.

McMillan, D. W., and David M. Chavis. 1986. Sense of Community: A Definition and Theory. *Journal of Community Psychology* 14 (1): 6–23.

Milofsky, Carl (ed.). 1988. *Community Organizations: Studies in Resource Mobilization and Exchange.* New York: Oxford University Press.

Ministry of Health (MoH). 1998. *Proceedings of the 3rd National Scientific Conference on Occupational Health.* Hanoi, December 1998.

Ministry of Planning and Investment (MPI). 1999. *Report of the Development Strategy Center.* Hanoi: MPI.

Mol, Arthur. 1995. *The Refinement of Production—Ecological Modernization Theory and the Chemical Industry.* Utrecht: International Books.

Mydans, Seth. 1996. Tiger Economy Has Become a Fading Vision For Vietnam. *New York Times,* July 25, Section D, p. 1.

Nader, Ralph and Lori Wallach. 1996. GATT, NAFTA, and the Subversion of the Democratic Process. In *The Case against the Global Economy,* ed. J. Mander and E. Goldsmith, 92–107. San Francisco: Sierra Club Books.

National Academy of Public Administration (NAPA). 2000. *environment.gov— Transforming Environmental Protection for the 21st Century.* Washington, D.C.: NAPA.

Niskanen, W. 1975. Politicians and Bureaucrats. *Journal of Law and Economics* 18: 617.

Nguyen Duc Khien. 1996. Report on the Environmental Management Practices of Hanoi's DOSTE. Paper presented at the VCEP Seminar on Environmental Management, Hanoi, August 20–23, 1996.

Nguyen, Hai. 1993. "Ministry Alarmed Over Surging Chemical Pollution." *Vietnam Investment Review* (September 13).

O'Connor, David. 1994. *Managing the Environment with Rapid Industrialization: Lessons from the East Asian Experience.* Development Centre of the Organization for Economic Co-Operation and Development, Paris: OECD.

O'Connor, James. 1988. Capitalism, Nature, Socialism: A Theoretical Introduction. *Capitalism, Nature, Socialism* 1 (1): 11–38.

O'Connor, James. 1994. Is Sustainable Capitalism Possible? In *Is Capitalism Sustainable? Political Economy and the Politics of Ecology,* ed. Martin O'Connor, 152–175. New York: Guilford Press.

O'Connor, James. 1998. *Natural Causes—Essays in Ecological Marxism.* New York: Guilford Press.

O'Connor, Martin. (ed.). 1994. *Is Capitalism Sustainable? Political Economy and the Politics of Ecology.* New York: Guilford Press.

Office of Technology Assessment (OTA). 1994. *Industry, Technology, and the Environment: Competitive Challenges and Business Opportunities.* Washington, D.C.: Government Printing Office.

Opschoor, Hans, and Kerry Turner. 1994. *Economic Incentives and Environmental Policies: Principles and Practice.* Boston: Kluwer Academic Publishers.

Organization for Economic Cooperation and Development (OECD). 1985. *Environmental Policy and Technical Change.* Paris: OECD.

O'Rourke, Dara. 1995. Economics, Environment and Equity: Policy Integration During Development in Vietnam. *Berkeley Planning Journal* 10: 15–35.

O'Rourke, Dara. 1997. Smoke from a Hired Gun: A Critique of Nike's Labor and Environmental Auditing in Vietnam as Performed by Ernst and Young. Unpublished manuscript, available on the Internet at: <www.corpwatch.org/trac/nike/ernst/>.

O'Rourke, Dara, Nghiem Ngoc Anh, Vu Manh Hai, Peter B. Evans, Pham Viet Hung, Nguyen Thi Lam, Dao Minh Truong, and Martha Kendall Winnacker. 1995. Environment and Industrial Renovation in Vietnam: A Report from Vinh Phu Province. Institute of International Studies Working Paper, Berkeley, Calif.: IIS.

O'Rourke, Dara, and Garrett Brown. 1999. Beginning to Just Do It: Current Workplace and Environmental Conditions at the Tae Kwang Vina Nike Shoe Factory in Vietnam. Unpublished manuscript, March 14, 1999, available on the Internet at: <www.globalexchange.org/economy/corporations/nike/vt.html>.

O'Rourke, Dara, Lloyd Connelly, and C. P. Koshland. 1996. Industrial Ecology: A Critical Review. *International Journal of Environment and Pollution,* 6 (2/3): 89–112.

O'Rourke, Dara, and Gregg Macey. 2003. Community Environmental Policing: Assessing New Strategies of Public Participation in Environmental Regulation. *Journal of Policy Analysis and Management,* 22, (3).

Ostrom, Elinor. 1990. *Governing the Commons: The Evolution of Institutions for Collective Action.* Cambridge: Cambridge University Press.

Ostrom, Elinor. 1996. Crossing the Great Divide: Coproduction, Synergy, and Development. *World Development* 24 (6): 1073–1087.

Pargal, Sheoli, H. Hettige, M. Singh, and D. Wheeler. 1997. Formal and Informal Regulation of Industrial Pollution. *World Bank Economic Review,* 11 (3): 433–450.

Pargal, Sheoli, and David Wheeler. 1995. *Informal Regulation of Industrial Pollution in Developing Countries: Evidence from Indonesia.* Policy Research Working Paper 1416. Washington, D.C.: World Bank.

Pearce, David, and Jeremy Warford. 1993. Planning Failure: Socialist Planning and the Environment. In *World Without End: Economics, Environment, and Sustainable Development,* ed. D. Pearce and J. Warford. New York: Oxford University Press.

Peltzman, S. 1976. Toward a More General Theory of Regulation. *Journal of Law and Economics* 19 (2): 211–240.

Pepper, David. 1993. *EcoSocialism—From Deep Ecology to Social Justice.* New York: Routledge.

Poffenberger, Mark. 1998. *Stewards of Vietnam's Upland Forests.* Berkeley, Calif.: Asia Forestry Network.

Porter, Gareth. 1993. *Vietnam: The Politics of Bureaucratic Socialism.* Ithaca: Cornell University Press.

Porter, Michael, and C. van der Linde. 1995. Green and Competitive: Ending the Stalemate. *Harvard Business Review* 73 (5): 120–134.

Portes, Alejandro. (ed.). 1995. *The Economic Sociology of Immigration: Essays on Networks, Ethnicity and Entrepreneurship.* New York: Russell Sage Foundation.

Portney, Paul. 1990. *Public Policies for Environmental Protection.* Washington, D.C.: Resources for the Future.

Prakash, Aseem. 2000. *Greening the Firm: The Politics of Corporate Environmentalism.* Cambridge: Cambridge University Press.

Princen, Thomas, and Matthias Finger. 1994. *Environmental NGOs in World Politics.* London: Routledge.

Putnam, Robert. 2000. *Bowling Alone: The Collapse and Revival of American Community.* New York: Simon & Schuster.

Putnam, Robert. 1993. *Making Democracy Work.* Princeton: Princeton University Press.

Rambo, A. Terry. 1994. Poverty, Population, Resources and Environment as Constraints on Vietnam's Development. Paper presented at the Eleventh Annual Berkeley Conference on Southeast Asian Studies, February 26–27.

Rawls, John. 1971. *A Theory of Justice.* Cambridge, Mass.: Belknap.

Riedel, James, and Bruce Comer. 1995. Transition to Market Economy in Viet Nam. Paper prepared for the Asia Foundation Project on Economies in Transition: Comparing Asia and Eastern Europe.

Risse-Kappen, Thomas. (ed.). 1995. *Bringing Transnational Relations Back In—Non-State Actors, Domestic Structures and International Institutions.* Cambridge: Cambridge University Press.

Rock, Michael. 2001. Pathways to Industrial Environmental Improvement in the East Asian Newly Industrializing Economies. Paper presented at the Ninth International Greening of Industry Conference, Bangkok, Thailand, January 21–25.

Rondinelli, Dennis. 2000. Rethinking US Environmental Protection Policy: Management Challenges for a New Administration. The PricewaterhouseCoopers Endowment for The Business of Government, grant report, November.

Roodman, David Malin. 1999a. Vietnam: The Paradox of Public Pressure. *Worldwatch Magazine* 12 (6) (November/December).

Roodman, David Malin. 1999b. Regulatory Framework for Industrial Pollution: Constraints and Opportunities. Paper presented at the Conference on Sustainable Industrial Development, Ministry of Industry, Ministry of Science, Technology, and Environment, and World Bank, Hanoi, June 10–11.

Ross, Lester. 1998. The Politics of Environmental Policy in the People's Republic of China. In *Ecological Policy and Politics in Developing Countries*, ed. U. Desai, 47–64. Albany: State University of New York Press.

Rueschemeyer, Dietrich, and Peter B. Evans. 1985. The State and Economic Transformation: Toward an Analysis of the Conditions Underlying Effective Intervention. In *Bringing the State Back In*, eds. P. B. Evans, D. Rueschemeyer, and T. Skocpol, 44–77. New York: Cambridge University Press.

Sabatier, Paul. 1999. *Theories of the Policy Process.* Boulder: Westview.

Sabel, Charles, Archon Fung, and Brad Karkkainen. 1999. Beyond Backyard Environmentalism. *Boston Review* 24 (5): 4–23.

Sam, Dinh Van. 1998. Environmental Management in Vietnam. Unpublished paper, Hanoi, Vietnam.

Sapru, R. K. 1998. Environmental Policy and Politics in India. In *Ecological Policy and Politics in Developing Countries*, ed. U. Desai, 155–182. Albany: State University of New York Press.

SarDesai, D. R. 1992. *Vietnam: The Struggle for National Identity.* Los Angeles: University of California Press.

Schnaiberg, Allan. 1994. The Political Economy of Environmental Problems and Policies: Consciousness, Conflict, and Control Capacity. *Advances in Human Ecology* (3): 23–64.

Scott, James. 1976. *The Moral Economy of the Peasant—Rebellion and Subsistence in Southeast Asia.* New Haven: Yale University Press.

Shaw, Randy. 1999. *Reclaiming America: Nike, Clean Air and the New National Activism.* Berkeley, Calif.: University of California Press.

Sikor, Thomas O., and Dara O'Rourke. 1995. A Tiger in Search of a New Path: The Economic and Environmental Dynamics of Reform in Viet Nam. *Vietnam's Socio-Economic Development* Hanoi, 4: 35–52.

Sikor, Thomas O., and Dara O'Rourke. 1996. Economic and Environmental Dynamics of Reform in Vietnam. *Asian Survey,* XXXVI (6): 601–617.

Socialist Republic of Vietnam (SRV). 1993. *Vietnam: A Development Perspective.* Report prepared for the Paris Donors Conference.

Socialist Republic of Vietnam (SRV). 1994. *Law on Environmental Protection.* Hanoi: National Political Publishing House.

Socialist Republic of Vietnam (SRV). 1995. *Viet Nam National Environmental Action Plan,* Hanoi, Draft April 20, 1995.

Spaargaren, G., and A. Mol. 1992. Sociology, Environment, and Modernity: Ecological Modernization as a Theory of Social Change. *Society and Natural Resources* 5: 323–344.

Stavins, R. N., and B. W. Whitehead. 1992. Market based incentives for environmental protection. *Environment* 34: 7–11, 29–42.

Stewart, Richard. 1993. Environmental Regulation and International Competitiveness. *The Yale Law Journal,* 102 (8): 2039–2106.

Stier, Ken. 1996. Northern Vietnam's Most Polluted City Shows Industry's Challenge. German Press Agency, May 4, 1996.

Stromseth, Jonathan. 1998. Reform and Response in Vietnam: State–Society Relations and the Changing Political Economy. Unpublished Ph.D. Dissertation, Columbia University, New York.

Szasz, Andrew. 1994. *EcoPopulism: Toxic Waste and the Movement for Environmental Justice.* Minneapolis: University of Minnesota Press.

Tarrow, Sidney. 1994. *Power in Movement—Social Movements, Collective Action and Politics.* Cambridge: Cambridge University Press.

Tarrow, Sidney. 1996. States and Opportunities: The Political Restructuring of Social Movements. In *Comparative Perspectives on Social Movements,* eds. D. McAdam, J. McCarthy, and M. Zald, 41–61. New York: Cambridge University Press.

Taylor, Bron Raymond. (ed.). 1995. *Ecological Resistance Movements: The Global Emergence of Radical and Popular Environmentalism.* Albany: State University of New York Press.

Than, Mya, and Joseph Tan. 1993. *Vietnam's Dilemmas and Options.* Singapore: Institute of Southeast Asian Studies.

Thanh, Nguyen Cong. 1993. *Vietnam Environment Sector Study.* Manila: Asian Development Bank.

Thomas, Robert. 1994. *What Machines Can't Do—Politics and Technology in the Industrial Enterprise.* Berkeley: University of California Press.

Thrift, Nigel, and Dean Forbes. 1986. *The Price of War: Urbanization in Vietnam, 1954–85.* Boston: Allen and Unwin.

Tietenberg, Tom, and David Wheeler. 1998. Empowering the Community: Information Strategies for Pollution Control. Paper presented to the conference on Frontiers of Environmental Economics, Arlie House, Virginia, October 23–25, 1998.

United Nations Development Program (UNDP). 1995. *Incorporating Environmental Considerations into Investment Decision-Making in Viet Nam.* Hanoi, December.

United Nations Development Program (UNDP). 1997. *Incorporating Environmental Considerations into Urban Planning.* Vietnam Capacity 21 Project, Hanoi, May.

United Nations Development Program (UNDP). 1999. *A Study on Aid to the Environment Sector.* Report submitted to the Government of Vietnam, Hanoi.

United Nations Industrial Development Organization (UNIDO). 1991. *Audit and Reduction Manual for Industrial Emissions and Wastes.* Technical Report No. 7, Vienna.

United Nations Industrial Development Organization (UNIDO). 1995. *Vietnam—Industrial Environmental Protection Policies,* UNIDO Project NC/VIE/94/020, Hanoi.

United Nations Industrial Development Organization (UNIDO). 1998. Incentives for Pollution Prevention and Control in Dong Nai. Report to the project VIE/95/053, August 6, 1998.

Vietnam Investment Review (VIR). Hanoi, Vietnam, multiple issues, including Nov. 17, 1997.

Vietnam News. 1999. Hanoi, Vietnam, Multiple issues.

Vogel, David. 1995. *Trading Up: Consumer and Environmental Regulation in a Global Economy.* Cambridge: Harvard University Press.

Vogel, David. 1999. Environmental Regulation and Economic Integration. Paper prepared for a workshop on Regulatory Competition and Economic Integration, Yale Center for Environmental Law and Policy, New Haven, October 1999.

Wade, Robert. 1990. *Governing the Market—Economic Theory and the Role of Government in East Asian Industrialization.* Princeton, N.J.: Princeton University Press.

Wallace, David. 1995. *Environmental Policy and Industrial Innovation: Strategies in Europe, the U.S., and Japan.* London: Earthscan.

Wang, Hua. 2000. Pollution Charge, Community Pressure and Abatement Cost: An Analysis of Chinese Industries. World Bank Policy Working Paper. Washington, D.C.: World Bank.

Wang, Hua, and David Wheeler. 1999. Endogenous Enforcement and Effectiveness of China's Pollution Levy System. World Bank Policy Working Paper. Washington, D.C.: World Bank.

Wapner, Paul. 1996. *Environmental Activism and World Civic Politics.* Albany: State University of New York Press.

Weale, Albert. 1992. *The New Politics of Pollution.* Manchester: Manchester University Press.

Wilkinson, K. P. 1986. In Search of the Community in the Changing Countryside. *Rural Sociology* 51 (1): 1–17.

Williams, H. E., J. Medhurst, and K. Drew. 1993. Corporate Strategies for a Sustainable Future. In *Environmental Strategies for Industry: International Perspectives on Research Needs and Policy Implications,* eds. K. Fischer and J. Schot, 117–146. Washington, D.C.: Island Press.

Woolcock, Michael. 1998. Social Capital and Economic Development: Towards a Theoretical Synthesis and Policy Framework. *Theory and Society* 27 (2): 151–208.

Woolcock, Michael, and Deepa Narayan. 1999. Social Capital: Implications for Development Theory, Research, and Policy. *World Bank Research Observer,* February, Washington, D.C.

World Bank. 1991. *Environmental Assessment Sourcebook.* Environment Department, Volumes I and III. Washington, D.C.: World Bank.

World Bank. 1993. *Viet Nam—Transition to the Market.* Washington, D.C.: World Bank.

World Bank. 1994. *Vietnam—Environmental Programs and Policy Priorities for an Economy in Transition.* Agriculture and Natural Resources Division, East Asia and Pacific Region, Washington, D.C.

World Bank. 1995. *Vietnam—Economic Report on Industrialization and Industrial Policy.* Country Operations Division, East Asia and Pacific Region, Washington, D.C.

World Bank. 1997. *Vietnam—Economic Sector Report on Industrial Pollution Prevention.* Agriculture and Environment Operations Division, Washington, D.C.

World Bank. 1999. *Greening Industry: New Roles for Communities, Markets and Governments.* NIPR. Washington, D.C.: World Bank.

Zald, Meyer. 1996. Culture, ideology, and strategic framing. In *Comparative Perspectives on Social Movements,* eds. D. McAdam, J. McCarthy, and M. Zald. New York: Cambridge University Press.

Index